Guide to
ENVIRONMENTAL LAWS

From Premanufacture to Disposal

Guide to ENVIRONMENTAL LAWS

From Premanufacture to Disposal

James A. Stimson
Jeffrey J. Kimmel
Sara Thurin Rollin

The Bureau of National Affairs, Inc., Washington, D.C.

Library of Congress Cataloging-in-Publication Data

Stimson, James A.
Guide to environmental laws : from premanufacture to disposal /
James A. Stimson
p. cm.
Includes index.
ISBN 0-87179-712-7
1. Hazardous substances—Safety measures. I. Kimmel, Jeffrey J.
II. Rollin, Sara Thurin. III. Title
T55.3.H3S75 1993
344.73′046—dc20 93-12472
[347.30446] CIP

Published by BNA Books
1250 23rd St., NW, Washington, D.C. 20037–1165

Printed in the United States of America
International Standard Book Number: 0–87179–712–7

PREFACE

This book uses a transactional approach to explain the functions and interrelationships of the environmental laws, regulations, and enforcement strategies that affect every phase of the manufacturing process. As environmental regulations become increasingly complex, it was felt that this approach would better serve the reader in gaining a more practical understanding of the law.

The laws, or in some cases major titles of the laws, are grouped into six sections that coincide with major phases of the manufacturing process: Premanufacture Controls, Right-To-Know Requirements, Manufacturing Controls, Transport and Identification, Storage, and Disposal and Cleanup.

Guide to Environmental Laws: From Premanufacture to Disposal begins with programs under the Toxic Substances Control Act and the Federal Insecticide, Fungicide, and Rodenticide Act, which tell companies what chemicals they may manufacture and distribute. The book proceeds to programs under the Clean Air Act and Clean Water Act, which regulate pollutant releases during the manufacturing process. There are separate chapters on waste transport and storage, concluding with waste disposal programs under the Resource Conservation and Recovery Act and the Comprehensive Environmental Response, Compensation, and Liability Act, better known as Superfund.

Although the organization of the book is sequential, the chapters are self-contained, so readers may go immediately to the chapters that interest them without reading the preceding material.

Most of the programs are operated by the Environmental Protection Agency, although the Department of Transportation plays a role in hazardous waste transport and in water pollution control activities of the U.S. Coast Guard. Several other agencies are involved in various capacities.

Readers of this book may find it useful to consult a companion BNA Books' publication, *U.S. Environmental Laws*, which contains the text of all the statutes discussed in the primer. Those interested in state activities may wish to consult BNA's *Guide to State Environmental Programs*.

I extend my thanks to the writers, editors, and others whose contributions helped make this book a reality.

JAMES A. STIMSON

March 1993

CONTENTS

Preface v

Introduction 1

Chapter 1. Premanufacture Controls

I. Toxic Substances Control Act 17
Overview 17
Chemical Reporting and Recordkeeping Requirements 19
General Reporting Rules 19
Preliminary Assessment Information Rule 20
Comprehensive Assessment Information Rule ... 21
Confidential Business Information 22
TSCA Inventory of Existing Substances 22
Inventory Updates 23
Allegations of Adverse Effects 24
Unpublished Health and Safety Studies 25
Substantial Risk Notices 25
Chemical Testing 26

Testing Agenda 26
Existing Chemical Testing 27
Test Rules 28
Negotiated Agreements 29
New Chemical Testing 30
Test Guidelines and Good Laboratory Practices 31
New Chemicals 32
Premanufacture Notification 33
PMN Exemptions 33
PMN Forms, Fees, Recordkeeping 34
Restrictions 35
Section 5(e) Orders 35
Significant New Use Rules (SNUR) 37
Biotechnology Products 39
Hazardous Chemicals: TSCA Section 6 39
Imminent Hazards 40
PCBs 40
Equipment Reclassification 41
Labeling 41
Fire Rules 41
Inspections, Recordkeeping 42
CFCs 42
Asbestos 43
Chromium 44
Metal-Working Fluids 44
Chemical Exports, Imports 45
Export Notice 45
Import Certification 46
Citizens' Suits, Petitions 47
Enforcement 48
Inspections, Subpoenas 48
Administrative Proceedings 49
Federal Court Proceedings 50
Penalty Policy 50
Examples 51

II. Federal Insecticide, Fungicide, and Rodenticide Act 52
Overview 52

Product Registration 55
  Registration Procedures 56
    Exemptions 57
    Exclusions 58
  Data Requirements 58
    Product Chemistry 59
    Coloration 59
    Data Requirement Tables 59
    Minor-Use Pesticides 61
  Data Call-In 61
  Pesticide Assessment Guidelines 62
  Good Laboratory Practices 63
  Meeting Data Requirements 64
    Citation of Data 65
    Data-Use Compensation 65
    Formulator's Exemption 66
    Exclusive Use 66
    Trade Secrets 67
    Public Access 67
  Fees 68
Classification 68
Applicator Certification 69
  Commercial Applicator Standards 70
  Private Applicator Standards 71
Product Labeling, Packaging 72
  Warning Statements 73
  Child-Resistant Packaging 73
Application Review Standards 74
  Conditional Review 75
  Unconditional Review 76
  Approval or Denial of Registration 77
Emergency Exemptions 78
General Reporting, Recordkeeping 80
  Establishment Registration 80
  Annual Report 81
  Producers' Recordkeeping 81
  Recordkeeping for Dealers of Restricted-Use Pesticides 82
  Recordkeeping for Experimental Use Permits 82

Adverse Effects Reporting 82
Substantial Risk Reporting 83
Tolerances, Experimental Use Permits 84
Tolerance Petition Process 85
Experimental Use Permits 86
Application Process 87
Reregistration of Older Pesticides 87
Five-Phase Plan 88
Application Process 89
Risky Pesticides 90
Special Review 90
Suspension 91
Cancellation 92
Voluntary Cancellation 92
Hearing Process 93
Indemnification 93
International Trade Restrictions 94
Imports 94
Exports 94
State Authority 95
Special Local Needs Registration 96
Enforcement 97
Violations 98
Penalties 99
State Primacy 100
Enforcement Examples 101
State Actions 101
Federal Administrative Proceedings 101
Federal Courts 103

Chapter 2. Right-to-Know Requirements

Introduction 105
Emergency Planning 106
State and Local Governments 107
Substances and Facilities Covered 107
Notification 108
Emergency Response Plans 108
Emergency Releases 109

Annual Chemical Inventories 112
MSDS Reporting 113
Inventory Reporting 114
Public Access to Reports 114
Toxic Chemical Release Reporting 115
Facilities Covered 115
Chemicals, Thresholds 116
Exemptions 117
Reporting 117
Pollution Prevention 119
Trade Secrets 119
Enforcement 121

Chapter 3. Manufacturing Controls

I. Clean Air Act 123
Introduction 123
National Ambient Air Quality Standards (NAAQS) .. 124
Carbon Monoxide 126
Lead 126
Nitrogen Dioxide 127
Ozone 127
Particulates 129
Sulfur Dioxide 130
New Source Performance Standards 130
Meeting the Standards 133
State Implementation Plans 133
The New Compliance Regime 135
Prevention of Significant Deterioration 137
Mobile Sources 137
Exhaust Emissions 139
Evaporative Emissions 141
Recalls 142
Acid Rain Control 143
The Problem 143
The Solution 145
Phase I 145
Phase II 146
Air Toxics 147

Hazardous Air Pollutants 149
Arsenic 150
Asbestos 150
Benzene 151
Beryllium 151
Mercury 152
Radionuclides 152
Radon-222 153
Vinyl Chloride 153
The New Air Toxics Regime 153
Permits 155
Enforcement 155
Summary 158

II. Clean Water Act 158
Introduction 158
The Permit System 159
What Are Navigable Waters? 160
Water Quality 161
Effluent Limitations 162
Industrial Sewage Treatment 164
BCT, BPT, and BAT 166
Municipal Sewage Treatment 167
Pretreatment Programs 168
State Toxics Programs 169
Wetlands 170
Nonpoint Source Pollution 172
Enforcement 175

III. Safe Drinking Water Act 179
Introduction 179
Wellhead Protection 180
Sole-Source Aquifers 180
Standards 181
Responsibilities of Drinking Water Suppliers 182
Underground Injection 184
Enforcement 184

Chapter 4. Transport and Identification

I. The Hazardous Materials Transportation Act 187
Introduction 187
Shippers 190
Hazard Classification 191
Hazardous Materials Table 192
Packaging, Containers 194
Performance-Oriented Packaging 194
General Packaging Requirement 195
Exceptions 195
Forbidden Materials, Packaging 196
Container Reuse 196
Miscellaneous Provisions 197
Identification, Labels, Placards 197
Identification Numbers 198
Labels 199
Placards 200
Shipping Papers 201
Shipping Certification 203
Hazardous Waste Manifest 204
Exemptions 204
Carriers 206
Mode-Specific Carrier Rules 206
Rail Transport 207
Air Transport 208
Water Transport 209
Highway Transport 210
Spills, Emergency Response 211
Written Report 213
DOT Emergency Response Guidebook 213
International Restrictions 214
United Nations 215
State Role 215
Enforcement 217
Examples of Administrative Proceedings 219
Federal Railroad Administration 219
Federal Aviation Administration 220

Federal Highway Administration 220
U.S. Coast Guard 221
RSPA 221
Examples of Federal Court Proceedings 222

II. Identification and Transport of Waste under RCRA 222
Introduction 222
Identification, Management of Solid Waste 224
Solid Waste 224
Exclusions 225
Nonhazardous Solid Waste 225
State Plans 226
Identification of Hazardous Waste 227
Hazardous Waste Characteristics 228
Hazardous Waste Lists 228
Delisting 229
Regulation of Generators 229
Small-Quantity Generators 230
Hazardous Waste Transport 231

Chapter 5. Storage

I. Regulation of Storage Tanks under RCRA 233
Introduction 233
Program Scope 234
Technical Requirements 235
Financial Requirements 237
State UST Programs 238
Other Provisions 239
Trust Fund 239
Enforcement 239

II. Hazardous Waste Storage under RCRA 239

III. Hazardous Material Storage under TSCA 240

IV. Pesticide Storage under FIFRA 242

V. EPCRA Notification Requirements 243

Chapter 6. Disposal and Cleanup

I. Disposal of Waste under RCRA 245
Introduction 245
Treatment, Storage, and Disposal Facilities 246
What Is a TSD Facility? 247
Interim Status Facilities 248
Final Permits 249
Corrective Action 251
Bans on Land Disposal of Waste 253
State Program Approval 254
Enforcement 255

II. CERCLA 258
Introduction 258
Site Selection 260
What Is a Hazardous Waste Site? 260
Hazard Identification 261
The Cleanup Process 262
Background 262
Site Studies 263
Remedies: Emergency, Long-Term 266
PRP Involvement in Remedy Selection 268
State Responsibilities 269
Court Review of Remedies 269
Enforcement 270
Background 270
Section 106: EPA Administrative Orders 271
Section 107: Going to Court 272
Elements of a Section 107 Enforcement Suit 273
Liable Parties 273
Release or Threatened Release 274
Response Costs 275
Standard of Liability 275
Going After Other PRPs 276
Defenses to Liability 277
The Most-Used Defense: Third-Party Defense .. 278

Settlement Authorities 279

III. Pesticide Disposal under FIFRA 282
Disposal of Cancelled or Suspended Products 283
Pesticide Containers 284

IV. Relationship between SDWA Standards and Disposal Laws 284

Appendix A: Air Quality 287
Appendix B: Clean Water Charts 311
Appendix C: Hazardous and Solid Waste Charts 327

Index 333

INTRODUCTION

The year 1970 was a watershed year for the environment, and not just because of Earth Day: the U.S. Congress and the executive branch began replacing an ad hoc system of environmental laws with a comprehensive system designed to regulate all pollution sources and all media.

On January 1, 1970, President Richard Nixon signed the National Environmental Policy Act (Public Law 91-190). It required the federal government to assess the environmental impact of its own activities and created the Council on Environmental Quality, an independent advisor to the White House. This panel was to play an important role in creating and upgrading the laws that protect our environment.

Also in 1970, Nixon issued an executive order consolidating various environmental programs into a single agency—the Environmental Protection Agency (EPA)—that was to administer the laws passed by Congress.

The Clean Air Amendments of 1970 (Public Law 91-604) established a system whereby all states would develop plans to meet federal air quality standards previously developed by the Department of Health, Education, and Welfare. In 1970, President Nixon also signed the Resource Recovery Act, which not only promoted recycling but launched an investigation of hazardous waste disposal practices that eventually led to passage of the Resource Conservation and Recovery Act (RCRA; Public Law 94-580) and the Toxic Substances Control Act (TSCA; Public Law 94-469), both enacted in 1976.

The Federal Water Pollution Control Act Amendments of 1972 (Public Law 92-500) required treatment of all municipal and industrial discharges into U.S. waters and greatly expanded federal aid for sewage treatment plant construction. The Federal Insecticide, Fungicide, and Rodenticide Act (FIFRA; Public Law 92-516), also vintage 1972, required testing and registration of pesticides.

Congress found, however, that simple regulation and enforcement was not enough to control pollution that was the result of decades of neglect. Contamination of the Love Canal area in Niagara Falls, New York, discovered in 1978, impelled Congress to pass in 1980 the Comprehensive Environmental Response, Compensation, and Liability Act (CERCLA; Public Law 96-510). This law created the Superfund, an emergency fund the government could use to clean up sites that pose immediate health threats. More significant, it established a liability system whereby those who caused the contamination could be billed for the cleanup.

Thus, the regulatory statutes fall into three broad categories:

- *Premanufacturing and Marketing Controls:* TSCA and FIFRA actually dictate what substances industries can produce and market. Under TSCA, any new substance must be evaluated for safety before it is mass produced. FIFRA requires registration of new pesticides and reregistration of existing pesticides.

- *Pollution Controls:* The clean air and water laws, along with RCRA, are traditional "end-of-pipe" control laws that regulate industrial and municipal releases to air, water, and soil. These systems are largely executed through permits that dictate either the concentration or total amount of pollutants that may be released into the environment. The statutes generally do not dictate what an industry may produce.

- *Liability Controls:* Finally, the Superfund law is unique among these statutes because its roots are in the legal system, not the regulatory system. Its liability scheme casts a broad net to find owners and past owners of polluting facilities to pay for Superfund cleanups. Its doctrine of strict joint and several liability is designed to hit the "deep-pocket" contributors—Superfund sites often contain waste from many companies—and force them to pay at least their fair share.

The authors' transactional approach to the subject matters details each point in the manufacturing cycle where the various authorities of the law come into play. Three broad divisions of statutory authority are used: premanufacturing and marketing controls—the development and testing stage; pollution controls—on releases to air, land, and water; and liability controls—on transport and disposal of waste.

PREMANUFACTURING AND MARKETING CONTROLS

Two environmental laws, TSCA and FIFRA, were designed to control the use of chemicals as products or components of products. EPA's mandate from these statutes is to balance the risks and benefits associated with the use of chemicals.

EPA weighs chemicals' potential to cause harm to human health or the environment against their potential to benefit society. Both TSCA and FIFRA contain an array of regulatory options, and EPA has the authority to ban, restrict, or control the use of chemicals under each.

In a 20th-anniversary issue of the *EPA Journal*, the agency's in-house publication, EPA stated that it has three goals in its regulation of chemicals in commerce: (1) to prevent chemicals from entering the market if they pose an unacceptable risk, (2) to manage the use of chemicals that are inherently risky so that society can enjoy their benefits safely, and (3) to remove chemicals from the market when the agency concludes they pose an unreasonable risk.

While TSCA and FIFRA share similar goals and have similar regulatory requirements, their evolution and ability to command public attention vary.

Toxic Substances Control Act

TSCA was enacted in 1976, after the other major pollution control laws were already on the books. The purpose of TSCA is to protect the public from chemicals that present a hazard to health or the environment. The law was designed to fill regulatory gaps left by the pollution control laws that set contamination limits for air, water, and land.

Because TSCA empowers EPA to regulate chemicals at their source, companies that make, import, export, or process chemicals are required to maintain records. The act's recordkeeping and reporting requirements are its most prominent feature.

EPA's TSCA authority extends to all chemicals in commerce. The agency's jurisdiction includes new chemicals as well as substances that have been in commerce for centuries, such as lead and mercury.

Before it can regulate a chemical, EPA must be able to prove that it presents a risk or that there is significant exposure.

TSCA has become identified with certain substances because parts of the statute direct EPA to impose limits on them. To some people, TSCA is known as "the PCB [polychlorinated biphenyl] statute," because most uses of PCBs were banned under TSCA.

Others maintain that EPA has yet to fully apply the law's authorities. Nonetheless, all companies that handle chemicals must be familiar with TSCA, especially its reporting requirements.

On the other hand, TSCA authority was delimited by a federal appeals court decision—and the Justice Department's refusal to challenge—that EPA's 1989 asbestos ban was invalid. The appeals court ruled that EPA failed to consider less burdensome alternatives to an outright ban of 94 percent of all uses (*Corrosion Proof Fittings v. EPA*, 947 F.2d 1201, 33 ER Cases 1961 (5th Cir. 1991)).

Some chemicals are exempt from TSCA regulation because they are covered by other federal laws. TSCA exemptions include food and food additives, drugs, cosmetics, tobacco, nuclear material, and pesticides.

Federal Insecticide, Fungicide, and Rodenticide Act

The first U.S. pesticide law was enacted in 1910. In the past seven decades, the law has been revised six times. FIFRA, as the law has been known since 1947, requires companies to register their pesticide products with EPA before they can be marketed or used in the United States. This licensing process affects pesticide manufacturers, formulators, importers, distributors, brokers, packagers, and users.

EPA evaluates new pesticide registrations to determine whether the chemical will present unreasonable adverse effects when used according to its labeling instructions. New pesticide products—those manufactured after November 1, 1984—must be reviewed by EPA before they can be marketed.

When the agency reviews a registration, it must weigh the product's risks and benefits. The statute requires EPA to consider the economic, social, and environmental costs and benefits when making a registration decision.

In 1988, FIFRA was amended to require EPA to review old pesticides, those products registered before November 1, 1984. This review process is called reregistration. The public outcry about the safety of two "existing" pesticides,

Alar and ethylene dibromide (EDB), is often credited with spurring Congress to revise the law. The safety of thousands of pesticides is expected to be reviewed by the scheduled close of reregistration in 1997.

In the 1990s, EPA is expected to apply its TSCA and FIFRA authorities to the review of products developed through biotechnology. Congress also is expected to reauthorize FIFRA. Major changes are anticipated in the areas of pesticide exports and food safety. There is no comparable congressional interest in TSCA, although an amendment to tighten control on lead is possible. Because TSCA does not have a built-in expiration date, there is no statutory need for it to be reauthorized.

POLLUTION CONTROLS

Three statutes impose limits on releases of pollutants to air, land, and water. The Clean Air Act authorizes the government to set air quality standards and impose emission limits on individual pollution sources. Major emitters now obtain permits. A separate program has been set up to control acid rain.

The Clean Water Act, in similar fashion, set in motion the writing of water quality standards and individual discharge limits. The National Pollutant Discharge Elimination System (NPDES) governs the issuance of discharge permits under the statute.

Finally, individuals depositing waste on or in the ground must obtain permits and meet various standards under RCRA. These standards are far more stringent for wastes defined as hazardous, a condition that has prompted a vigorous and ongoing debate about the definitions of hazardous and nonhazardous waste.

All three programs authorize the government to take administrative action against alleged violators or take them to court if necessary. States play a major role, because EPA's general policy is to delegate to states the authority to run the programs on a day-to-day basis. Such delegation typically occurs when the state enacts a statute or adopts an administrative program that is at least as effective as the federal program. After the delegation of authority, oversight becomes EPA's primary role. States also have been delegated much responsibility to run the Safe Drinking Water Act program, which consists of protecting underground and above-ground sources of drinking water.

Clean Air Act

EPA has five principal duties under the Clean Air Act:

- Setting air quality standards
- Setting emission limits
- Regulating motor vehicle emissions
- Controlling acid rain
- Enforcement

The agency sets air quality standards for six common air pollutants: carbon monoxide, nitrogen dioxide, sulfur dioxide, ozone, lead, and particulates. EPA also is setting standards for a large number of hazardous air pollutants. If a city or county violates an air quality standard, the state is obligated to develop a State Implementation Plan (SIP) that will bring about compliance with that standard.

Emission limits can take several forms: They can represent rates, such as one pound of carbon monoxide per hour. They can also consist of specific technologies that have been proven effective in limiting a pollutant; most new standards for hazardous air pollutants are expected to take this form. Finally, they can take narrative form, as do measures of particulate effects on visibility.

Regulation of motor vehicles consists of limits on emission rates, usually expressed in grams per mile. The limits cover not only tailpipe emissions but evaporative emissions.

Evaporative emissions largely occur during refueling and just after the engine is turned off, but so-called running losses add to the problem.

The acid rain program is essentially the imposition of sulfur dioxide emission limits on utilities that burn coal. Even though most U.S. utilities do not cause violations of air quality standards on a local level, their cumulative sulfur dioxide and nitrogen oxide emissions can travel to other regions. Acid rain is blamed for acidifying soil and water, which in turn threatens plants and wildlife.

EPA has several types of enforcement authority under the Clean Air Act. The agency can issue administrative orders and impose administrative penalties. It also can threaten states with the loss of certain federal funds if they fail to adopt an acceptable SIP. EPA can impose a Federal Implementation Plan if the SIP process stalls. The agency also can take alleged violators to court in both civil and criminal proceedings.

New authority to require permits for major pollution sources will give EPA a new type of charge in administrative, civil, and criminal action: violation of permit conditions. When permit conditions are met, however, EPA may have more difficulty convincing a court to order remedies or impose penalties.

Clean Water Act

The Clean Water Act relies more than any other program on its permit scheme (NPDES). Most states have been delegated authority to issue NPDES permits under the system. Virtually all discharges to all water bodies are prohibited unless they are covered by an NPDES permit.

EPA still has a duty to set water quality standards and discharge limits, although states can adopt their own standards if they convince EPA that they are sufficiently protective of human health and the environment.

NPDES permits cover only discharges to water bodies. Discharges to public sewer systems are covered by EPA's pretreatment program. Any factory that sends wastewater to a municipal sewage treatment plant can be required to pretreat its waste by the local sewer authorities. In general, industrial wastewater may not contain contaminants that disrupt the sewage treatment plant or pass through it unchanged.

Wetlands are considered U.S. waters and, as such, are subject to the Clean Water Act's prohibitions against unpermitted discharges. EPA's permitting authority is limited to oversight. The U.S. Army Corps of Engineers has the primary permitting responsibility for activities involving wetlands.

States are required to identify waters in which toxic contamination is interfering with their desired use. States then must identify the pollutant, find its source, and impose appropriate controls.

Municipalities and private companies also must be covered by permits for stormwater pollution. Such pollution is not discharged directly, but washes off the land during rainstorms. Soil at construction sites, grease and oil on city streets, and pesticides on farmland all find their way into lakes and streams during rainfall. EPA's plan is to cover most parties with general permits, which only require that broad standards be met. Some parties will need to obtain individual or group discharge permits for stormwater, based on the nature of the activity or discharge.

A key to EPA's enforcement strategy under the Clean Water Act is its national municipal policy, which states that all public sewage treatment plants that do not have adequate treatment capability must be put on a court-approved schedule to attain that capacity. The agency also is focusing on state and local pretreatment programs.

As with most pollution control laws, EPA has the authority to issue administrative orders and impose administrative penalities, or may take violators to court.

Safe Drinking Water Act

Public water supplies are protected by the Safe Drinking Water Act in two ways. The first is through standards set for various pollutants. The second is by protection of water supplies, mostly porous rock formations known as aquifers.

The setting of maximum contaminant levels for drinking water largely parallels procedures for other media, such as the air quality standards set under the Clean Air Act and the water quality standards set under the Clean Water Act.

The wellhead protection program established by the Safe Drinking Water Act is essentially a pollution-prevention program. States are authorized to designate wellhead protection areas, identify pollution sources, and impose appropriate controls.

States also may ask EPA to designate sole-source aquifers, defined as a population's sole or principal source of drinking water. Once such a designation is approved, no federal funds may support any activity that would contaminate that aquifer.

Drinking water suppliers, meanwhile, face a significant pollution-prevention challenge of their own: keeping lead out of public water supplies. Their task is complicated by the fact that some older housing has lead in its plumbing, which introdues the contaminant at the point of use.

Hazardous Materials Transportation Act

Transportation of chemicals and hazardous materials is regulated under HMTA. The law, enacted in 1975 and revised in 1990, gives the Department of Transportation (DOT) authority to regulate materials that may pose an unreasonable risk to health, safety, property, or the environment during transport by air, water, highway, and rail.

Shippers, brokers, carriers, transporters, and those who make, repair, or test materials for hazardous materials transportation must register with DOT. The agency classifies materials by hazard. DOT may regulate any safety aspect of the transportation process, such as packing, repacking, handling, labeling, marking, placarding, and routing. These types of requirements are contained in DOT's Hazardous Materials Table.

HMTA empowers the federal government to assess civil and criminal penalities for violations. Enforcement and compliance for hazardous materials transportation is shared among federal agencies. When materials are transported by only one mode, that modal agency enforces HMTA. When materials are shipped by more than one mode, DOT's Research and Special Programs Administration enforces the law.

Resource Conservation and Recovery Act

By 1976, Congress was ready to tackle the growing problem of trash disposal in the United States. That year RCRA was passed, with an emphasis on safe disposal of solid waste and encouragement for recycling. However, by reauthorization in 1984, problems with hazardous waste disposal had taken center stage.

There are now three major aspects of the law: (1) control of solid waste, (2) control of hazardous waste, and (3) regulation of underground storage tanks.

- *Solid Waste:* The federal system of regulation is minimal, with the brunt of responsibility falling on states. The most stringent requirements are imposed on operators of municipal landfills, and few duties are imposed on industrial sites.
- *Hazardous Waste:* This aspect of regulation received the most attention in the 1984 RCRA amendments, with Congress spelling out in unusual detail the steps that EPA was required to take in regulating generators, transporters, and facilities that handle hazardous waste. Simple land disposal of such waste is prohibited in almost all cases, and the treatment technologies demanded by EPA can be quite complex. In addition, anyone who operates a facility that treats, stores, or disposes of hazardous waste must take extreme precautions to ensure that the site does not become a candidate for Superfund cleanup later.
- *Underground Storage Tanks:* Again in the 1984 amendments, Congress set out a system to track and respond to

leakage from the estimated 3 million underground tanks in the United States. In addition to reporting requirements, EPA and state authorities must devise technical standards to prevent further problems with new storage tanks. Lastly, Congress set up a cleanup fund, similar to the Superfund for abandoned hazardous waste sites.

LIABILITY CONTROLS

When Congress passed CERCLA in 1980, the goal was to fund cleanups of abandoned hazardous waste sites and force the parties responsible for the conditions to pay. The law consisted of two main components: (1) a Superfund of money for waste site cleanups, along with a plan for carrying out remedies, and (2) a liability system allowing the federal and state governments to recover site response costs.

Under the law:

- EPA, with state help, can select and rank hazardous waste sites across the United States that are in need of attention, placing the most hazardous sites on the National Priorities List.
- These sites are cleaned up according to the general principles set out by EPA in the National Contingency Plan (NCP), which provides guidelines for remedial investigations and feasibility studies to be conducted prior to cleanup. The NCP also provides standards for determining the appropriate level of waste cleanup and specifying the extent to which states can participate in cleanup decisions.
- EPA can use money in the trust fund established under CERCLA—the Superfund—to clean up the sites. Then, along with the Justice Department, the agency can use enforcement authorities granted in the law to sue parties responsible for site conditions to recover site response costs.
- The agency can also issue administrative orders to responsible parties, requiring them to carry out site clean-

ups and threatening severe penalties if they ignore the orders.

- States are given authority to help EPA determine the appropriate remedy for a waste site, participate in cleanup funding, and sue to recover their cleanup costs.

The most contentious aspect of the law is the liability standard applied to parties who have been linked to a hazardous waste site. When the government goes after such parties for response costs, the court reviewing the suit can apply strict joint and several liability—meaning that a single party can be held responsible for millions of dollars of response costs, no matter how much that party was responsible for site conditions or how many other parties were involved at the site.

Emergency Planning and Community Right-to-Know Act

Along with the 1986 amendments to CERCLA, Congress created a separate law called the Emergency Planning and Community Right-to-Know Act (EPCRA). After several tragic incidents around the world involving chemical spills, including the notorious release in Bhopal, India, in 1984 that killed more than 2,500 people, public pressure came to bear on EPA and Congress to respond.

There are four basic sets of requirements under EPCRA:

- *Emergency Planning:* State and local governments must establish a framework for handling emergency releases and potential accidents. Industry has input into the planning, and industrial facilities are required to notify local officials if they handle certain chemicals.
- *Emergency Releases:* If there is a release of any one of over 700 chemicals defined as hazardous under the Superfund law, the responsible facility must report the release if the amount exceeds a given reportable quantity.
- *Chemical Inventory:* Facilities are required to submit specific information on chemicals handled at the plant, culled

from material safety data sheets (as already required by the Occupational Safety and Health Act) and from EPA reporting forms.

- *Routine Toxic Releases:* The most controversial part of the new law requires facilities to provide an annual list of releases and amounts of regulated chemicals. This information is then included in a computerized toxic release inventory database that is accessible by the public.

Sources of Environmental Law: Common Law

Until the early 1970s, there were few avenues available for citizens and the federal and state governments to respond to pollution problems. Although there were scattered laws stating general policies in favor of a clean environment, virtually none of the legislation had the "teeth" to force substantial changes in industry practices.

Prior to the birth of the environmental movement in the United States, remedies for pollution came from legal principles developed for centuries under English and American law. When a person felt that he or she had been damaged by another's polluting activities, the usual course was to go to court. The victim would file suit, using any of the following claims:

- *Nuisance:* This involves an action interfering with a person's use or enjoyment of his or her property. For example, if smoke from a neighboring factory makes a person's property uninhabitable, that person could sue that factory owner for creating a nuisance.
- *Trespass:* Trespass occurs when a person or object has entered another's property without permission. A person entering a property and dumping garbage would provide the basis for a trepass suit by the property owner.
- *Negligence:* There are four elements to a negligence claim: (1) a duty (2) that has been breached (3) that caused (4) harm to another. For example, the owner of a hazardous waste site that leaks and contaminants the drinking water wells of neighbors could be sued for negligence. The site owner had a duty to safely store waste; this duty was

breached by the leakage. This, in turn, caused harm to the neighbors by preventing them from using a drinking water source.

The cumbersome process of going to court and proving these claims, along with the high costs of litigation, helped spur the move toward pollution control legislation.

Enforcement

Federal and state regulators need more than regulations to get companies to clean up the environment. The "stick" wielded to force compliance is the enforcement system authorized under each environmental statute. Through an active administrative, civil, and criminal enforcement program, federal and state agencies send a message that the private sector will have a hard time profiting from ignoring environmental requirements.

Normally, there are three levels of enforcement, in ascending order of severity: (1) administrative, (2) civil, and (3) criminal.

- *Administrative Enforcement:* Normally, an agency such as EPA or a state equivalent can issue an order to an alleged violator requiring it to take specific action to remedy compliance problems. In addition, the agency may sometimes impose penalties. The alleged violator is usually entitled to a hearing before the agency to contest the charges or negotiate compliance terms.
- *Civil Enforcement:* If violations are serious enough, or the alleged violator has not complied with administrative orders, the federal or state agency can go to court to seek an injunction and/or civil penalties. Usually, such suits are filed by the attorney general's office on behalf of the regulatory agency. At this stage, penalties tend to be higher than in administrative actions, and violation of a court-ordered remedy can be penalized in court through a citation for contempt of court.
- *Criminal Enforcement:* For the worst violators, federal and state regulators have the option of having criminal

charges filed in court by the attorney general's office. Criminal charges are usually associated with instances where a person or company officer knew that his or her actions were violating an environmental law. In addition to increased monetary fines, a violator will face potential jail time.

Courts

Courts play two roles under federal environmental laws: (1) as a check on the actions of regulatory agencies and (2) as a forum for enforcement suits. Whenever an agency such as EPA issues a final order or takes other final action, parties who are affected by the action can go to court to challenge it. In enforcement cases, the court will determine whether violations have occurred and the appropriate remedy.

In the federal system, there are three levels of courts:

- U.S. district courts, which can review enforcement actions and some types of regulatory challenges. Consisting of single judges, these trial courts handle most environmental litigation.
- U.S. appeals courts, which usually have three judges who hear appeals from district court rulings and review the majority of challenges to agency action.
- U.S. Supreme Court, the highest court in the United States. The nine justices rarely grant reviews of environmental cases. In a typical year, perhaps four environmental disputes are reviewed. Both types of cases—regulatory challenges and enforcement cases—are presented to the Supreme Court.

State court systems are usually based on a parallel structure, with a trial court, intermediate appellate court, and state supreme court. From the state supreme court, cases can be taken to the U.S. Supreme Court, although the high court rarely reviews such cases.

Chapter 1

PREMANUFACTURE CONTROLS

I. TOXIC SUBSTANCES CONTROL ACT

OVERVIEW

Congress enacted the Toxic Substances Control Act (TSCA) in 1976, after six years of debate, in response to environmental damage from two types of chemicals: polychlorinated biphenyls (PCBs) and chlorofluorocarbons (CFCs). TSCA gives the Environmental Protection Agency (EPA) authority to regulate the manufacture and distribution of virtually all industrial chemicals.

Most provisions of TSCA apply to companies that make, import, process, or distribute chemicals in commerce. The law does not prohibit the manufacture or use of dangerous or toxic chemicals. Under TSCA, EPA can regulate chemicals if the substance presents an "unreasonable risk," or EPA has reason to believe the substance may present an unreasonable risk to human health or the environment. Risk is defined as the probability of an occurrence of an unwanted, adverse effect.

Congress instructed EPA to balance the benefits of the use of a chemical with its potential to cause injury. The agency must avoid regulations that would "impede unduly or create unnecessary economic barriers to technological innovation."

The statute provides the agency with a range of actions it can take to control risks from chemicals. Because risk cannot be measured in uniform terms, federal officials make regulatory decisions based on an assessment of available information. Accordingly, TSCA was designed to enable EPA to gather information about chemicals as well as control risks associated with chemicals. Under TSCA, EPA can:

- Require companies to test certain chemicals identified as possible risks (Section 4)
- Impose restrictions on the manufacture of a new substance or a "significant new use" of a chemical (new or existing) if EPA determines the compound may pose a risk or there would be significant exposure to the compound (Section 5)
- Ban or control the manufacture, processing, use, disposal, or distribution of a substance that presents an "unreasonable risk" (Section 6) or "imminent hazard" (Section 7) to health or the environment
- Require companies to report information on production volume, use, and effects on health and the environment (Section 8)

Congress, recognizing that international trade is a source of chemicals in commerce, added provisions in TSCA for monitoring the import and export of chemicals. Section 12 of the act requires companies to notify EPA of exports, and Section 13 requires importers to certify that the chemical complies with TSCA.

The law authorizes the federal government to take civil and criminal actions to enforce TSCA (Section 15). TSCA gives EPA broad authority to conduct inspections and issue subpoenas (Section 11). EPA also has authority to assess civil fines (Section 16). The maximum fine is $25,000 per day.

Unlike most other environmental laws, TSCA does not provide for state enforcement.

CHEMICAL REPORTING AND RECORDKEEPING REQUIREMENTS

EPA's primary source of information for chemical regulations is that collected under TSCA's reporting and recordkeeping requirements. EPA uses these data to make hazard and safety decisions about chemicals. TSCA contains five types of chemical reporting requirements:

- Section 8(a) provides for general production and use reporting rules.
- Section 8(b) requires EPA to compile and maintain the TSCA Inventory of Existing Substances.
- Section 8(c) requires firms to maintain records of alleged significant adverse reactions.
- Section 8(d) enables EPA to collect unpublished health and safety studies conducted by companies.
- Section 8(e) requires anyone who obtains information indicating that a chemical may pose a substantial risk to human health or the environment to notify EPA within 15 days.

General Reporting Rules

EPA uses Section 8(a) of TSCA to require chemical makers, importers, and some processors to report data on production, use, disposal methods, by-products, and human exposure (40 CFR 704). Companies are required to keep copies of their TSCA reports and supporting data for three years (40 CFR 704.11). There are three types of reporting rules issued under this section of the law: the preliminary assessment information rule (PAIR), the comprehensive assessment information rule (CAIR), and chemical-specific rules.

PAIR and CAIR, also known as generic reporting rules, were set up to streamline information gathering under the law. Both rules use standardized forms with a uniform set of questions. To get PAIR- and CAIR-type information, EPA simply amends the model rule with a list of chemicals.

PAIR generally is designed to provide data to support EPA's chemical testing program under TSCA. CAIR information gathering is geared toward the agency's risk assessment and risk management activities.

After EPA developed the two generic rules, it stopped using chemical-specific reporting rules to collect information. These rules were too resource intensive for EPA to write because each one was unique. However, in the first 15 years of TSCA, the agency issued nine chemical-specific rules (40 CFR 704, Subpart B).

Preliminary Assessment Information Rule

PAIR is designed to gather information to support the agency's health and environmental assessment efforts. In particular, the data are used to support TSCA test rules. Since the rule took effect in 1982, EPA has required companies to provide information on more than 250 chemicals (40 CFR 712).

Manufacturers and importers of PAIR-listed chemicals must report production, use, and exposure information using EPA Form 7710-35. The two-part report contains questions about the manufacture of the chemical and what happens to the chemical once it reaches the customer or processor.

Generally, companies must report information from the latest complete corporate fiscal year. A company need not file a PAIR report if it sent EPA the same information within the past three years.

Exemptions are available for small companies, production of less than 500 kilograms (1,100 pounds), research-and-development production, and generation of a listed substance as a by-product or impurity in manufacturing (40 CFR 712.25).

EPA once sought a fine of $28,900 from a company for filing a PAIR report three months late. The fine was reduced by half in negotiations (*In re* Merichem Co. TSCA 91-H-05, agreement approved June 29, 1991).

Comprehensive Assessment Information Rule

CAIR was designed to standardize the process of gathering detailed information from chemical makers, importers, and processors. The model rule was issued in 1988. The questions on the CAIR form enable EPA to obtain information necessary to assess risks posed by chemical substances and regulate as needed.

The reporting form contains 10 sections, each with a specific set of questions. Information on a chemical's production volume, market value, production costs, worker exposure, waste treatment, and environmental releases must be reported on the CAIR form. To date, 19 chemicals and numerous trade-name products containing those chemicals have been subject to CAIR (40 CFR 704, Subparts C and D).

When EPA adds a substance to the rule's coverage, it identifies specific sections of the form for reporting. EPA also designates what types of companies must file a report and the reporting period.

For example, in 1989 EPA required the makers of chlorine (CAS No. 7782-50-5) to complete a CAIR report. The agency also could have required importers and producers of the chemical to complete the report, but did not. The agency identified questions in 5 of the 10 sections for reporting and instructed chlorine makers to report information on the chemical from the period February 8, 1987, to February 5, 1989. The CAIR reports on chlorine were due to EPA within 90 days of the effective date of the rule.

CAIR reporting requirements may be waived for repackagers, small processors, and those who reported the same information on a complete corporate fiscal year within three years of EPA's request (40 CFR 704.210).

Confidential Business Information

Provisions for protecting trade secrets and confidential business information are an essential part of TSCA. Most of TSCA's reporting requirements allow for companies to claim information as confidential business information, which means that only EPA and its authorized contractors can have access to the data. A company's ability to claim data as confidential prohibits competitors from obtaining trade secrets while enabling EPA to review complete information about chemicals.

Congress wrote Section 14 of TSCA, disclosure of data, to guide EPA when it receives requests or petitions for the release of confidential information. This section of the law outlines how EPA should proceed if it determines that the release of the information is necessary. The agency's rules governing the handling of claims of confidentiality are located at 40 CFR 2.306.

TSCA Inventory of Existing Substances

EPA is required by TSCA Section 8(b) to compile, keep current, and publish a list of all chemical substances manufactured, imported, or processed for commercial purposes in the United States. About 78,000 chemicals were on the TSCA inventory as of January 1, 1993 (40 CFR 710).

The inventory is at the center of TSCA's regulatory framework. Chemical regulations are divided into two types: new chemical control and existing chemical control. If a chemical is on the TSCA inventory, it is an existing chemical. Once the initial TSCA inventory was published in May 1979, any chemical not included was considered a new chemical substance and was subject to premanufacture notification and review (40 CFR 710.2(r)). The premarket review process for new chemicals is addressed later in this chapter.

Because companies need to protect confidential business information reported to the inventory, the agency maintains two versions—a public inventory and a confidential, or CBI, inventory. The agency also maintains a master inventory file, a comprehensive list of chemicals that includes those recently added.

To determine whether a chemical is a new chemical, a company may ask the agency to search the CBI inventory or the master inventory file. EPA grants only requests that meet certain criteria (40 CFR 720.25).

The printed version of the inventory is contained in more than a dozen paperback volumes. Copies can be purchased from the U.S. Government Printing Office. The inventory is also sold on computer tape by the National Technical Information Service. In addition, there are several commercial databases that sell and service the inventory.

Inventory Updates

When Congress created the TSCA inventory, it mandated that the list of chemicals in commerce be updated periodically. In 1986, EPA amended the inventory reporting rules to require reporting every four years, beginning in 1986 (40 CFR 710.23). Records documenting inventory reports must be retained for four years beginning with the effective date of the reporting period (40 CFR 710.37).

Companies that make or import 10,000 pounds (4,540 kilograms) or more of a substance on the inventory must report the substance during the recurring reporting periods. Data on the production volume, plant site, and site-limited status of the substance during a corporate fiscal year must be reported for each plant site. A site-limited chemical is a substance that is made and used at the same plant. Partial updating occurred in 1986 and 1990.

Reporting can be done in writing or with magnetic media (i.e., computer tape, diskettes). For written reporting, companies must use EPA Form 7740-8 ("Partial Updating of TSCA Inventory Data Base Production and Site Report"). Copies of the form are available from EPA. Certain magnetic media were approved for reporting in 1990 (40 CFR 710.32).

Polymers, inorganic substances, microorganisms, and naturally occurring substances are generally excluded from TSCA inventory updating. Certain small manufacturers, companies that manufacture or import small quantities solely for research and development purposes, are among the firms not subject to this reporting requirement (40 CFR 710.29, 710.30).

Allegations of Adverse Effects

Under TSCA Section 8(c), companies are required to keep records of allegations of significant adverse reactions to any substance or mixture they manufacture, import, process, or distribute. The law requires companies to keep records and reports of such allegations for 30 years. EPA issued a rule in 1983 outlining the recordkeeping requirements, reporting criteria, and exemptions (40 CFR 717).

Companies must record any allegations that a substance caused irreversible health damage, such as cancer, or caused significant harm to the environment (i.e., fish kills).

Companies need not record known human health effects described in scientific articles or the firm's product labeling, environmental contamination already reported to the federal government, or anonymous allegations.

EPA may require the reporting of allegations on a chemical-specific basis through a TSCA Section 8(c) record call-in notice in the *Federal Register*. The agency does so when it needs information from a large number of companies or when the identities of potential respondents are unknown.

For example, in January 1988 EPA issued a TSCA Section 8(c) record call-in. The agency ordered manufacturers, importers, processors, and distributors of chemicals in the categories tri(alkyl/alkoxy) phosphates and diisocyanates to submit records of allegations of adverse effects. The call-in listed 51 chemicals in all.

If the agency seeks information only from a few known companies, it can contact them directly by letter. The reporting period will be included in the notice or letter. Companies also must permit inspection of records by an EPA representative.

Unpublished Health and Safety Studies

Section 8(d) of TSCA authorizes EPA to require companies to report unpublished health and safety studies. EPA developed a model rule, similar to the PAIR and CAIR rules, for collecting these studies (40 CFR 716).

The agency collects unpublished health and safety data when it is deciding whether to issue a TSCA test rule. Chemicals are added to the reporting rule twice a year following the Interagency Testing Committee (ITC) recommendations for testing under TSCA. The committee is composed of officials from throughout the federal government. The role and membership of the committee are discussed in the Testing Agenda section of this chapter.

The reporting period for each chemical lasts for 10 years from the time the chemical is covered by the model rule. The reporting requirement applies to current and prospective companies that make, import, or process the listed substance. As of 1991, there were several hundred chemicals listed on the TSCA Section 8(d) rule.

Substantial Risk Notices

Chemical manufacturers, distributors, processors, and importers must inform EPA under TSCA Section 8(e) if they obtain new information that supports a conclusion that a substance or mixture presents a substantial risk of injury to health or the environment. Information reported to the agency under other EPA-administered statutes, such as the Clean Air Act, need not be reported on a TSCA substantial risk notice.

The reporting requirement took effect at the same time TSCA became effective (January 1, 1977). By 1990, EPA had received nearly 1,500 Section 8(e) substantial risk notices, and the agency has since published several volumes of such notices.

In March 1978, EPA issued a policy statement explaining its interpretation and enforcement plans for Section 8(e) of the law. To clarify how this part of the law works, the agency issued question-and-answer documents in March 1978 and October 1988 and a reporting guide in June 1991. Additional guidance is anticipated. EPA has not issued formal regulations on Section 8(e).

The substantial risk reporting requirement applies to all companies subject to TSCA. There are no exemptions for small business or for research and development. Companies that obtain information about serious toxic effects from a chemical (i.e., death, cancer, birth defects, neurotoxicity, and ecological effects) must report that information to EPA within 15 days.

CHEMICAL TESTING

Reporting requirements are not the only way EPA gathers chemical information. The chemical testing program established by TSCA is the agency's second route for getting information about chemicals. The agency can require chemical manufacturers to conduct studies on substances and mixtures already on the market (Section 4) or about to come on the market (Section 5). The agency can require companies to conduct studies to determine health effects of a substance, its chemical fate, and its environmental effects. These data are used to determine whether and how to regulate potentially harmful chemicals.

Testing Agenda

EPA's testing agenda comes from many sources. Because the number of chemicals in commerce was significant when Congress wrote TSCA, the lawmakers established ITC to recommend to EPA chemicals that should receive priority attention under TSCA for Section 4 testing. The committee, composed of representatives from numerous federal agencies, including EPA, makes recommendations to EPA twice a year. These recommendations constitute the TSCA Section 4(e) priority testing list.

Program offices within EPA, such as the Office of Solid Waste or Office of Drinking Water, also recommend chemicals for TSCA testing. In addition, citizens' petitions filed under Section 21 of TSCA can be used to spur testing. Substances also are identified for testing through EPA's chemical screening activities and substantial risk information reported under Section 8(e) of TSCA.

TSCA's new chemical review program, as described in Section 5, is another source of information on chemicals (40 CFR 720). In this program, premanufacture notices (PMNs) are screened to identify chemicals of concern. Although TSCA does not mandate that testing be performed on new chemicals prior to commercialization, EPA can prohibit or limit the manufacture of a new substance pending testing if certain conditions are met. EPA must conclude that the new chemical may pose an unreasonable risk or that it will be produced in significant quantities.

Existing Chemical Testing

Under this section of TSCA, EPA can require companies to test existing chemicals in commerce. The agency can require testing when:

- The manufacture, distribution, use, or disposal of a chemical may pose an unreasonable risk to health or the environment and data are insufficient to support an evaluation of the risks associated with the substance (TSCA Section 4(a)(1)(A))
- The substance is produced in substantial quantities or there is significant exposure to the chemical or mixture (TSCA Section 4(a)(1)(B))

EPA can require industry to conduct the testing by issuing a test rule or negotiating an enforceable testing agreement. In either situation, EPA specifies whether chemical makers, importers, and/or processors must sponsor the testing. The agency established a standard schedule for developing test rules and negotiated agreements. It would take about a year to negotiate a testing agreement and about two years to develop a test rule, based on the EPA schedule (40 CFR 790).

Test Rules

Between 1984 and 1990, EPA issued about 30 chemical-specific test rules (40 CFR 799). Like other formal EPA rule-makings, these test rules are first proposed for public comment.

In a test rule, EPA recommends procedures for conducting studies, specifies when the results and final reports must be sent to the agency, and specifies the purity of the test chemical. After a review of public comment on the proposed rule, the rule is made final. Any subsequent modifications to a final test rule must be approved by EPA.

Because only one set of test data is needed to fulfill the obligations of a TSCA test rule, companies that must sponsor the testing often hire an independent laboratory and divide the test costs. If companies cannot agree to a cost-sharing arrangement, EPA can determine a fair allocation, but the agency has never had to use this authority (40 CFR 791).

In 1989, EPA issued a test rule for isopropanol (CAS No. 67-63-0). Because the chemical, also known as rubbing alcohol, is used by millions of people, the agency based its test rule on the occurrence of significant human exposure.

ITC had designated isopropanol for priority testing two years earlier. The required testing was estimated to cost the four producers of the chemical up to $3.8 million. EPA required the companies to conduct a comprehensive set of health effects studies on isopropanol because more than 22 million pounds of the chemical are used annually to make rubbing alcohol. Annual U.S. isopropanol production has exceeded 1 billion pounds per year since 1956, according to EPA. The agency cited its authority under TSCA Section 4(a)(1)(B), the potential for significant human exposure, when it issued the rule.

Negotiated Agreements

In 1986, EPA developed an alternative to test rules: enforceable testing agreements. These consent agreements are as binding as a test rule. Between 1986 and 1990, EPA and chemical companies signed nine enforceable agreements.

To initiate the negotiation process, EPA publishes a notice in the *Federal Register* inviting interested parties to participate in negotiations. Chemical makers, importers, processors, and other interested parties can participate. The negotiating sessions are open to the public.

If no consensus is reached after 10 weeks, EPA will stop negotiating and go to rule-making. If, however, an accord is reached, the test sponsors and EPA officials sign an enforceable consent agreement. Testing must begin after an announcement summarizing the agreement is published in the *Federal Register*.

In 1989, EPA and the Eastman Kodak Co. reached a fairly typical agreement—after 17 months of negotiations—on a series of tests to study the chemical fate and environmental effects of crotonaldehyde (CAS No. 4170-30-3). The chemical, which was included in two 1988 ITC reports to EPA, is used primarily as an intermediate in the production of crotonic acid, ascorbic acid, and other substances. Kodak, the sole U.S. producer of the chemical, agreed to do a study on aquatic biodegradation and a series of five acute toxicity tests. Depending on the results of those tests, additional studies may be triggered. The total estimated cost of the testing plan was not revealed under the terms of the negotiated agreement. EPA published an announcement describing the agreement on November 9, 1989.

New Chemical Testing

EPA can require a maker of a new chemical to conduct tests on the chemical before it is commercialized. Under TSCA Section 5, companies must notify EPA at least 90 days before marketing a new chemical. This advance notice, which is made by filing a PMN, gives EPA the opportunity to evaluate the potential risks associated with the manufacture of a new chemical and decide whether to put restrictions in place, including testing requirements. The premarket review process for new chemicals is addressed later in this chapter (40 CFR 720).

In the absence of sufficient information, EPA can require a company to conduct tests on a new chemical if it concludes that:

- The manufacture, processing, distribution in commerce, use, or disposal of the chemical may present an unreasonable risk of injury to health or the environment
- The substance will be produced in substantial quantities and may be reasonably anticipated to enter the environment in substantial quantities or there may be significant or substantial human exposure to the substance

When one of these two conditions is met, the agency and the chemical maker or importer may negotiate a TSCA Section 5(e) consent order. Usually, these consent orders place restrictions on the manufacture, use, or disposal of the compound. These restrictions apply only to the parties that sign the agreement.

As of October 1990, 145 of 522 TSCA Section 5(e) orders contained testing requirements. Usually, a company is permitted to make a certan amount of the new substance to help defray the costs of testing.

Because companies are not required to complete health and environmental effects studies as part of the PMN review process, EPA issued a policy statement in 1981 to encourage the voluntary development of such data. This policy was modeled on international efforts of the Organization for

Economic Cooperation and Development. In Europe, companies that want to commercialize a new chemical must complete a battery of studies called the Minimum Premarket Data Set. The policy statement includes a list of recommended tests and methods for evaluating potential effects of a new chemical.

Test Guidelines and Good Laboratory Practices

Testing required under TSCA must be conducted according to guidelines. Failure to follow these general instructions or EPA-approved substitute methods can result in enforcement action against the test sponsors and the laboratories performing the tests.

The agency included three types of test guidelines in the Code of Federal Regulations following litigation that required all chemical testing conducted under TSCA to be enforceable. The three TSCA test guidelines, which are continually being revised, are as follows:

- Chemical fate testing guidelines (40 CFR 796) with specific instructions about how to conduct studies to determine a substance's physical and chemical properties, transport characteristics, and transformation processes (i.e., biodegradability)
- Environmental effects testing guidelines (40 CFR 797) with procedures for the many studies used to determine the effects of a chemical on aquatic and terrestrial life
- Health effects testing guidelines (40 CFR 798) with procedures for short-term and long-term studies

In the early 1980s, EPA also issued a set of procedures called good laboratory practices (GLPs) to ensure the quality and integrity of test data developed under TSCA (40 CFR 792). GLPs include requirements for animal care, test facilities, personnel, and equipment maintenance. They also cover planning, monitoring, recording, and storing the data generated from a study. Upon completion of TSCA studies, companies must send EPA the data and a statement certifying that the tests were conducted in accordance with GLPs.

GLPs requires a testing facility to keep raw data and specimens for 10 years. After EPA audits the study's data, the test facility is notified that some of the specimens and samples may be discarded.

Violation of testing requirements can bring enforcement action against the companies that hire a laboratory and the laboratory itself. In 1990, four chemical importers and a testing laboratory were charged with failing to meet the testing requirements of a TSCA test rule for a substance in the dioxin/furan family. The four importers faced the same charges relating to alleged failure to follow a test rule, while the laboratory was charged with allegedly not following EPA's good laboratory practices. The laboratory negotiated a settlement that included a 7 percent "good faith" reduction and a fine of $14,000 (*In re* Triangle Laboratory Inc., TSCA 90-H-14, agreement approved Oct. 18, 1990).

NEW CHEMICALS

Section 5 of TSCA empowers EPA to assess the safety of new chemicals before their manufacture or import into the United States. The law requires companies to file a PMN with EPA at least 90 days before beginning such action. After the review period, a firm may make or import the new chemical if EPA has not taken action to ban or restrict its manufacture, use, or disposal.

If EPA determines that a certain use of a new chemical may cause an unreasonable risk, it can ban the chemical under TSCA Section 5(f), issue or negotiate a TSCA Section 5(e) consent agreement pending the development of data, or issue a TSCA Section 5(a)(2) significant new use rule (SNUR). To date, EPA has issued only four Section 5(f) bans.

Premanufacture Notification

Under TSCA, a chemical is considered a "new chemical" if it does not appear on the TSCA Inventory of Existing Chemical Substances. After EPA published the first inventory of chemicals in commerce in 1979, all companies intending to make or import chemicals not on the TSCA inventory had to comply with the agency's premarket review rules (40 CFR 700–721). There are a series of exemptions to the premarket notification process (40 CFR 723).

EPA evaluates about 1,200 PMNs and 300 PMN exemption requests each year. Most notices are approved within the 90-day review period. A processing fee of $2,500 must accompany each PMN or PMN exemption request. TSCA requires EPA to publish certain non-confidential information about each PMN or exemption request it receives.

To determine whether a substance is an existing chemical or needs to have a PMN filed, the TSCA inventory or the master inventory file is consulted. Because the confidential versions of the inventory and master file are not available to the public, companies can ask EPA to conduct a search. The agency will verify whether a substance is on the TSCA inventory only when a company establishes a "bona fide intent to manufacture" by sending EPA information about the substance.

After determining that the substance is indeed a new TSCA chemical and not subject to regulation under federal pesticide, food, or drug laws, a company should determine whether the substance qualifies for an exemption from the PMN process.

PMN Exemptions

The 90-day premarket review applies to all new chemicals, except those substances that are not regulated by TSCA and those that qualify for an exemption under TSCA Section 5(h)(2). Even though the term "exemption" is used, companies must file a scaled-down version of a PMN to obtain an exemption. Generally, EPA reviews exemption notices within 21 days (40 CFR 700).

The PMN review exemptions generally are granted for:

- Chemicals produced in low volume (1,000 kilograms or less) (40 CFR 723)
- Chemicals used in instant photography (the so-called Polaroid exemption) (40 CFR 723)
- Polymers that meet EPA criteria (40 CFR 723)
- Chemicals used in small quantities for research and development (40 CFR 720.36)
- Chemical mixtures
- Chemicals used only in pesticides, tobacco products, nuclear materials, guns and ammunition, foods, food additives, drugs, or cosmetics
- Chemicals used only for test marketing (40 CFR 720.38)
- Chemicals whose generation is incidental to manufacturing, including impurities, by-products, and non-isolated intermediates (40 CFR 720.30(h))
- Chemicals imported in articles or manufactured items (40 CFR 720.22(b)(1))
- Chemicals manufactured solely for export (40 CFR 721.3)

EPA can declare a substance ineligible for an exemption or deny the request if it determines the compound may threaten health or the environment. Firms that are granted PMN exemptions must keep records of annual production volume and other documentation for at least five years.

PMN Forms, Fees, Recordkeeping

PMN forms may be filed with the agency on behalf of more than one company, and records must be kept for five years.

Once EPA receives a completed EPA form 7710-25 and the fee, it determines whether the PMN review is required, assigns the chemical a PMN number, and sends the company a letter of acknowledgment. The letter includes the PMN number and the date the 90-day review period begins.

Fees are required for PMNs for new chemicals, exemption applications and notices, and significant new use notices. The agency does not charge for processing low-volume exemption notices and test market exemptions applications. Most processing fees are $2,500.

If the review period ends and EPA has not taken action to regulate the chemical, the firm may make or import the substance. Within 30 days of importing or producing a new chemical, the company must file a notice of commencement with EPA. In the first 12 years of the program, the agency received about 6,500 notices.

After the premanufacture review process is complete and the company that filed the PMN notifies EPA that manufacturing has begun, the agency automatically adds the chemical to the TSCA inventory of existing chemicals. Once the chemical is on the TSCA inventory, companies are free to make or import the substance unless EPA imposed restrictions—through a significant new use rule (SNUR)—on it during the PMN review process.

Restrictions

TSCA gives EPA two ways to restrict the manufacture, use, and disposal of new chemicals. The first tool EPA has is the TSCA Section 5(e) order. The second is the SNUR.

Section 5(e) Orders

A Section 5(e) order is used when EPA lacks adequate information to evaluate the risks associated with a chemical. These orders can be issued unilaterally by EPA or following negotiations with the chemical maker or importer. Of the 522 TSCA Section 5(e) orders issued during the first 11 years of the PMN program, EPA issued 16 unilaterally.

EPA can initiate a TSCA Section 5(e) proceeding if it concludes that:

- The manufacture, processing, use, distribution in commerce, or disposal of the chemical "may present an unreasonable risk of injury to health or the environment"

- The substance will be produced in "substantial quantities" and may be reasonably anticipated to enter the environment in substantial quantities or there may be significant or substantial human exposure

Section 5(e) orders can be used, for example, to require studies on health or environmental effects of a PMN chemical; limit production levels; require the company to notify EPA of uses other than those anticipated and described in the PMN; prohibit or limit the chemical's release into water; require the chemical maker to inform workers of potential hazards associated with the substance; and require that workers wear protective clothing during manufacturing, handling, use, or disposal of the substance.

If the Section 5(e) order is negotiated, the restrictions apply only to the parties that signed the order.

Negotiated consent orders often allow a company to make or import a certain amount of the chemical before restrictions are imposed. EPA commonly includes testing triggers in negotiated Section 5(e) orders, requiring testing once production or imports exceed a certain amount. (See page 29 for more on negotiated agreements.)

EPA settled its first enforcement action under TSCA Section 5(e) in January 1991 through negotiations with American Cyanamid Co. The company agreed to pay a $28,350 penalty and issue a company directive on the proper way to seek a change to a Section 5(e) order. EPA charged the firm with violating a 1983 Section 5(e) consent order that placed disposal conditions on wastes containing the new chemical. The agency based its case, which was filed in 1990, on a 1984 inspection and review of the company's records (*In re* American Cyanamid Co., TSCA 90-H-15, Jan. 23, 1991).

Significant New Use Rules (SNUR)

The second tool TSCA provides for restricting new uses of chemicals is the SNUR. Section 5(a)(2) of the law authorizes EPA to identify hazardous chemicals and require manufacturers, importers, and processors to provide advance notice to the agency of new uses of those chemicals. EPA uses this authority for both new and existing chemicals (40 CFR 721).

An SNUR takes effect as soon as it is proposed. Without this provision, firms could defeat the purpose of the SNUR by initiating uses before it takes effect.

Companies must file a significant new use notice at least 90 days before commencing the manufacture, import, or processing of a chemical for a new use. The PMN form is used for this purpose.

Once EPA receives a new use notice, it can take regulatory action to control the chemical's use under sections 5(e), 5(f), 6, or 7 of TSCA. These provisions empower EPA to place restrictions on chemicals in commerce. If the agency does not take regulatory action, it must explain its decision in a *Federal Register* notice.

Issuance of an SNUR essentially states EPA's position that a chemical is hazardous. For that reason, companies tend to opt for substitute chemicals that have proven readily available instead of filing new use notices. No new use notices have ever been filed.

Any company that files a significant new use notice must keep documentation on the notice for five years (40 CFR 721.40).

The agency has developed two methods for issuing SNURs: a chemical-by-chemical approach and an expedited approach. Historically, EPA has used the chemical-by-chemical approach for existing chemical SNURs. The expedited approach, which embodies the so-called generic SNUR, was designed to address new chemicals.

Existing Chemical SNURs. Between 1985 and 1990, EPA issued 16 existing chemical SNURs affecting 24 chemicals.

Although these new use rules were developed on a chemical-by-chemical basis, often more than one chemical was listed in a rule. These rules are often called "dead chemical SNURs" because most, if not all, of the existing uses of the chemicals have ceased.

One of these rules required companies to notify EPA if they intended to initiate any use of methyl n-butyl ketone (MBK) (CAS No. 591-78-6). In the mid-1970s, MBK was linked to neurological problems in workers at a treated fabrics plant in Ohio. The chemical was used as a solvent for lacquers, in lacquer and varnish removers, and as a solvent for oils, fats, and waxes. The SNUR required that anyone wishing to make, import, or process MBK for any use notify EPA at least 90 days in advance.

However, by the time EPA issued the SNUR for MBK in 1987, manufacturers had long since ceased to produce or import the chemical. No one in the United States has manufactured MBK since 1979.

New Chemical SNURs. In 1989, EPA created an expedited process for issuing new use rules and the "generic SNUR" after negotiating with industry and others. The agency can issue generic SNURs in three ways. The first involves using the standard process of issuing a proposed rule, receiving public comment, and issuing a final rule. The other two procedures also provide for public comment, but they have shorter timetables. In these two cases, unless EPA receives unfavorable comments, the SNURs automatically take effect (40 CFR 721 Subpart D).

One of the objectives of the revised process was to close a loophole in coverage of TSCA Section 5(e) consent orders, which apply only to companies that sign the negotiated orders.

The generic SNUR is used to extend the restrictions in a TSCA Section 5(e) order to all companies making a chemical. These SNURs also are called expedited SNURs and follow-up SNURs. EPA issues these rules as often as necessary to keep up with TSCA Section 5(e) orders. In 1990, the first year of the expedited SNUR, the agency issued such rules for more than 100 substances.

Biotechnology Products

EPA has been developing a proposed rule that would require companies that deal with microorganisms to complete the premarket review process. The agency says it will develop procedures for oversight of certain new organisms manufactured for general commercial use. EPA also has said its forthcoming biotechnology rules will include special notification procedures for biotechnology products that would be used only for research and development purposes.

HAZARDOUS CHEMICALS: TSCA SECTION 6

Congress created TSCA to empower EPA to control risks associated with extremely hazardous chemicals. The government's failure to prevent environmental degradation caused by CFCs and PCBs was identified by many congressional observers as an impetus for the passage of TSCA. Although these chemicals were used widely without any immediately apparent adverse effects, both classes of chemicals can cause long-lasting environmental problems.

EPA can regulate existing chemicals and mixtures under TSCA Section 6 if it determines the substance presents or will present an unreasonable risk to human health or the environment. Less than one year after TSCA became effective, EPA issued a final regulation outlining its procedures for issuing Section 6 rules (40 CFR 750).

Between 1977 and 1990, EPA completed 22 regulatory actions to control six chemicals using TSCA Section 6: PCBs, CFCs, asbestos, dioxins, chromium, and metal-working fluids. Of these rules, 15 addressed PCBs.

TSCA Section 6 gives EPA a broad range of control options, including authority to ban manufacture, processing, or distribution of a substance; prohibit certain uses; or require special labeling or disposal for a chemical. The statute, however, requires EPA to impose the least burdensome requirements to achieve the desired control.

When deciding whether to regulate a chemical under Section 6, EPA must weigh the potential benefits and risks from using the chemical against the potential economic impact of its regulation. Economic analysis and risk analysis are used to assess the impact of proposed regulation. TSCA mandates that EPA assess the availability of substitutes when deciding whether to regulate a chemical under this section of the act.

The agency must also determine whether TSCA, as compared with other federal statutes, is the most efficient means for controlling the risk. TSCA, after all, was written to fill regulatory gaps, not usurp existing regulatory authorities. When EPA determines that another part of the federal government is better suited to regulate a risk associated with a chemical, EPA can make a formal referral to another agency under TSCA Section 9 instead of taking regulatory action under Section 6. Most TSCA Section 6 rules are codified through the traditional notice and comment process, except when an imminent hazard exists.

Imminent Hazards

There are two means for controlling imminent hazards from existing chemicals under TSCA. When EPA proposes a Section 6 rule, it can specify that it take effect on publication. The agency, under TSCA Section 7, also can ask a federal district court to restrict an imminently hazardous chemical or mixture or authorize the agency to seize the chemical. This provision was designed to protect human health and the environment until a TSCA Section 6 rule takes effect. As of 1990, EPA had not used this authority.

PCBs

In Section 6(e) of TSCA, Congress mandated that EPA issue a series of rules to ban PCBs, require labeling and recordkeeping, and regulate all aspects of their handling and disposal. These chemicals were used in electrical equipment and in the hydraulic fluid of some heavy machinery. Between 1979 and 1990, EPA issued more than a dozen PCB-related rules (40 CFR 761).

As of July 1, 1979, most manufacture, use, and distribution of PCBs was prohibited by EPA. PCBs at concentrations of less than 50 parts per million may be imported or exported for the purposes of disposal under the prohibition. EPA also wrote a comprehensive set of PCB storage and disposal rules.

The agency authorized the continued use of PCBs when they are used in a totally enclosed manner, recycled, inadvertently released in an excluded manufacturing process in very low concentrations, or other authorized nonenclosed instances. EPA reports that the life expectancy of some enclosed electrical equipment may be well into the 21st century. One of the authorized nonenclosed uses is servicing of contaminated transformers.

Equipment Reclassification

PCB-containing transformers can be reclassified as PCB-contaminated electrical equipment or non-PCB transformers by draining, refilling, or other such servicing aimed at reducing the concentration of the chemical. All PCBs removed from transformers must be disposed of according to EPA rules.

Labeling

TSCA rules require PCB-containing items to be labeled during their use, service, storage, and disposal. The agency has a standard label with a warning statement in large lettering. Responsibility for labeling falls on people who handle the items.

Fire Rules

EPA also established a number of requirements to prevent PCB fires. High-temperature PCB fires can produce dioxins and thus create another set of hazards. The agency required that transformers containing PCBs have special protections. Local fire departments must be notified of the location of the equipment. Combustible materials, including paints, solvents, plastics, sawn wood, cardboard boxes, and paper, cannot be stored near PCB-containing equipment.

Inspections, Recordkeeping

The TSCA rules require owners of PCB-containing equipment to inspect and maintain records describing the maintenance and condition of the equipment. Transformers, for example, must be visually inspected at least once every three months. A reduced inspection schedule can be established if measures are taken to reduce the risks associated with the equipment's use.

CFCs

CFCs were once widely used as spray-can propellants, refrigerants, and cleaning solvents. This class of chemicals proved harmless to humans, but mounting evidence indicates that CFCs deplete the stratospheric ozone layer, which protects the earth from the sun's harmful rays. The observed effect is a thinning, or "hole," in the ozone layer above the Antarctic. A similar but less pronounced effect has been observed over the Arctic.

In 1980 EPA, along with other federal agencies, banned several uses of CFCs as aerosol propellants (40 CFR 762). The agency issued a rule under TSCA Section 6 restricting the manufacture, processing, and distribution of CFCs. The agency also prohibited the import or export of the substance for use as a propellant. EPA issued exemptions for certain essential uses, including the use of CFCs as diamond-grit spray and in some plastics. Companies that make CFCs must file annual reports with EPA certifying that their uses complied with the TSCA ban.

New evidence of environmental damage to the ozone layer prompted an international effort to curb all uses of

CFCs and related bromine compounds. Under the 1987 Montreal Protocol on Substances that Deplete the Ozone Layer, dozens of countries agreed to phase out the production of CFCs. This action addressed the CFC uses that were not banned by the TSCA rule (e.g., as solvents and refrigerants). In 1990, Congress incorporated CFC phase-out language into the Clean Air Act.

Asbestos

Ten years after EPA announced its intention to regulate asbestos under TSCA, the agency, in July 1989, issued a ban under Section 6. The agency banned in stages almost all manufacture, import, distribution, and processing of asbestos. EPA concluded that asbestos poses an unreasonable risk to human health. Many respiratory ailments, including lung cancer, are linked to exposure to friable (crumbling) asbestos (40 CFR 763).

EPA requires that asbestos-containing products that are subject to this rule be labeled. The agency exempts products from the ban if the manufacturer can demonstrate the the use will not result in an unreasonable risk to human health. In the first two years of the ban, EPA granted a few requests to use asbestos in glass products.

The effective dates of the staged intervals for the ban on manufacturing, importing, and processing asbestos are August of 1990, 1993, and 1996. EPA restrictions on commercial distribution of asbestos products take effect in August 1992, 1994, and 1997.

EPA issued a worker protection rule in 1987 to extend the workplace standard issued by the Occupational Safety and Health Administration. The rule provides protection to state and local government workers who are not covered by the Occupational Safety and Health Administration's rules (40 CFR 763, Subpart E, Appendix B).

A U.S. appeals court overturned the asbestos ban and phase-out rule on October 18, 1991. The court said EPA failed to identify the least burdensome regulatory alternative, failed to allow public comment on the methodology used to identify the benefits of the rule, and failed to adequately evaluate the toxicity of likely substitutes. (*Corrosion Proof Fittings v. EPA*, 947 F.2d 1201, 33 ER Cases 1961 (5th Cir. 1991)). The Justice Department told EPA in early 1992 that it would not ask the Supreme Court to review the ruling. The agency is now investigating other ways to regulate asbestos.

Chromium

In 1990, EPA banned the use of hexavalent chromium (CAS No. 10588-01-9) in certain air conditioning systems because of the potential for adverse human health effects from air emissions. EPA determined that exposure to hexavalent chromium is associated with an increased risk of lung cancer. Chromium compounds were used to prevent corrosion within cooling systems. The TSCA rule did not ban all uses of the compound. In fact, the rule specifically allows for their continued use in "industrial cooling towers and closed cooling water systems." In these cases, the equipment must be labeled to warn people of the increased risk of lung cancer when the chemical is inhaled. EPA also required companies distributing these chemicals to maintain records for two years (40 CFR 749).

Metal-Working Fluids

Four new chemicals designed to control corrosion in metal-working fluids were banned in 1984. Because the combination of these chemicals produced nitrosamines, a known human carcinogen, EPA took swift action to limit their use. The agency simultaneously proposed the TSCA Section 6 rule and issued an identical order under TSCA Section 5(f) because it was immediately effective (40 CFR 747).

These rules require distributors of the chemicals to notify customers of the rule's restrictions prior to shipment.

The rules also require that warning labels be placed on metal-working fluids containing the substances.

CHEMICAL EXPORTS, IMPORTS

Section 12 of the law requires companies to notify EPA when they export or intend to export chemicals regulated under TSCA. Companies must send EPA an annual notice identifying the countries where the chemical was sent.

Section 13 requires importers to certify that an imported substance is in compliance with TSCA. The import reporting requirements apply to all incoming chemical shipments, while the export reporting requirements only apply to some shipments (40 CFR 707).

TSCA imposes other obligations on chemical importers. The statute includes importers in its definition of a chemical manufacturer. Importers are responsible for conducting studies mandated by TSCA Section 4 test rules as well as several reporting and recordkeeping requirements under TSCA Section 8.

Export Notice

The export rules require a company to send EPA an annual notice listing each country to which the firm intends to ship regulated chemicals. The agency then notifies the importing country's ambassador or another designated official (40 CFR 707, Subpart D).

Companies must notify EPA if an exported substance is the subject of a TSCA Section 4 test rule or negotiated testing agreement, a new chemical control order issued under Section 5(e), new chemical testing requirements under Section 5(b), an SNUR issued under Section 5(a)(2), or certain TSCA Section 6 rules.

EPA has developed special guidelines for reporting exports of asbestos and PCBs. Anyone who intends to export PCBs or PCB-containing articles for any purpose other than disposal must notify EPA. In the case of asbestos, EPA must be notified when raw or bulk shipments are made and when certain asbestos-containing products are shipped.

Import Certification

The import provisions of TSCA are enforced by the U.S. Customs Service, which is part of the Department of Treasury. In December 1983, just before the import rules became effective, EPA issued a policy statement on chemical imports and TSCA that is codified at 40 CFR 707.20.

Import certification takes place at the point of entry into the United States. So-called "blanket" certifications may be requested from Customs to cover multiple shipments of the same substance over a one-year period (19 CFR 12.118–12.127, 127.88).

Since January 1984, importers or their designated official (a broker) have been required to sign the following statement for each import of chemicals subject to TSCA: "I certify that all chemical substances in this shipment comply with all applicable rules or orders under TSCA and that I am not offering chemical substances for entry in violation of TSCA or any applicable rule or order under TSCA."

This certification requires importers to determine whether the chemical is subject to section 5, 6, or 7 of TSCA. In particular, importers need to determine whether the chemical is a new or existing chemical by EPA's definition. If the substance is not on the TSCA inventory of existing chemicals, it may need to be reviewed in the premarket review process. (See the discussion on the new chemical review process earlier in this chapter.)

Certification also is required for imports of chemicals not subject to TSCA, such as pesticides or food additives. In this situation, importers must certify that compliance with TSCA is not required. This type of certification is commonly called a "negative TSCA certification." The statement reads: "I certify that all chemicals in this shipment are not subject to TSCA."

EPA and Customs will refuse entry to any shipment until the certification is properly filed. Customs also will detain a shipment if there are reasonable grounds to believe that it violates TSCA. Such shipments must be brought into compliance, exported, destroyed, or abandoned within certain prescribed time periods.

CITIZENS' SUITS, PETITIONS

Citizens can bring legal actions under TSCA through lawsuits and agency petitions. The statute also provides for judicial review of EPA rule-making.

Individuals can initiate citizens' suits to enforce certain provisions of TSCA or compel EPA to take actions required by the statute. Section 20 of the law requires that the citizen provide EPA notice of intent to file suit. The procedures for filing a citizens' suit are codified at 40 CFR 702.

The second route available for citizens to initiate TSCA action is a Section 21 petition, also known as a citizens' petition. These petitions are used to get EPA to initiate, amend, or repeal a rule addressing chemical testing (Section 4), reporting (Section 8), or hazardous substances (Section 6). In 1985, EPA issued a guidance document to help people prepare Section 21 petitions. Between 1978 and 1990, EPA received about 41 petitions and amendments to petitions.

Citizens' petitions have been filed on a variety of issues, including the definition of PCBs, health effects studies on methanol fuels, and regulation of chemicals emitted from carpeting.

EPA must either grant or deny the petition within 90 days of receiving it. To evaluate the merits of the petition, EPA may hold a hearing or conduct an investigation. If the petition is granted, EPA must promptly initiate a proceeding. If EPA denies the petition, the denial must be published in the *Federal Register*.

Within 60 days of a denial or the expiration of the EPA review period, the petitioner may ask a district court to review EPA's response. This is called a de novo review because the court reviews the entire petition. Several Section 21 petition responses have been reviewed by the courts.

Finally, Section 19 allows for judicial review of a TSCA regulation. There is a specified time period, after EPA publishes a final rule and before it becomes effective, during which such lawsuits may be filed.

ENFORCEMENT

Section 11 of TSCA gives EPA broad authority to conduct inspections and issue subpoenas. Unlike most environmental laws, TSCA does not provide for state enforcement. Cases are developed and litigated by officials at EPA headquarters and regional offices.

EPA can file TSCA lawsuits in two forums: administrative proceedings within EPA or litigation in U.S. district court. Six EPA administrative law judges (ALJs) handle all the agency's administrative proceedings, including those for alleged violations of TSCA. Allegations of criminal violations under TSCA can be litigated only in U.S. district courts. Penalties can be levied in both forums.

Inspections, Subpoenas

Inspections for possible TSCA violations originate in EPA regional offices or the National Enforcement Investigation Center in Colorado. A "duly designated" EPA official may inspect a company's records (excluding financial, sales, research, and personnel information), processes, controls, and facilities. Inspectors often look for violations of more than one environmental law.

The agency also can issue subpoenas to search for possible violations. EPA can use TSCA's subpoena authority to require the testimony of witnesses, make inquiries, obtain documents, and gather other information it deems necessary for enforcement. For example, in 1990 EPA issued subpoenas to about two dozen firms, searching for possible violations of the substantial risk reporting requirements under TSCA Section 8(e). To date, no enforcement actions have been brought as a result of TSCA Section 8(e) subpoenas.

EPA also can use its TSCA subpoena authority to investigate non-TSCA compliance issues. A U.S. court of appeals upheld this use in a 1988 decision. (*EPA v. Alyeska Pipeline Service Co.*, 836 F.2d 443, 26 ER Cases 2129 (9th Cir. 1988)).

Administrative Proceedings

The federal governmental files about 400 TSCA cases a year for administrative review. These enforcement actions must be conducted in an adjudicatory proceeding that is consistent with the Administrative Procedure Act (5 USC 554). Companies can pursue settlement before an administrative law judge or negotiate a settlement with EPA directly.

Negotiated settlements—consent agreements—are made enforceable through accompanying consent orders. These orders must be approved by EPA's Environmental Appeals Board. The board, which replaced the agency's chief judicial officer in 1992, operates independently of the ALJs.

Cases that are not settled by negotiations are presented at a hearing before an ALJ. After the hearing the judge issues a decision, which may include a fine. These rulings may be appealed.

The first level of appeal is to the Environmental Appeals Board. The process of appealing the board's decision depends on whether EPA or the defendant is seeking the appeal. The defendant can file an appeal in a U.S. appeals court, while EPA can only ask the board to reconsider the case. EPA cannot appeal the board's decision in federal courts.

In February 1992, EPA Chief Judicial Officer Ronald McCallum ruled that there is no statute of limitations for administrative actions under TSCA. This backed the decisions of several ALJs and resolved, at least for the time being, a challenge by industry (*In re* Minnesota Mining & Manufacturing Co., TSCA Appeal 90-3, Feb. 28, 1992). McCallum is now one of the three judges on the new Environmental Appeals Board.

Federal Court Proceedings

Although EPA investigates and develops both civil and criminal complaints, all allegations of criminal wrongdoing are litigated by government attorneys from the Department of Justice in U.S. district courts. The federal government has seldom used this authority.

In a criminal proceeding, the defendant has the right to a trial before a jury. Congress limited criminal prosecution under TSCA to misdemeanors, for which an individual can be imprisoned for up to one year. No felony charges are authorized under TSCA.

Penalty Policy

Congress established a civil penalty ceiling of $25,000 per day under TSCA Section 16. Each day a violation occurs constitutes a separate violation. In 1980, EPA developed a policy for calculating civil penalties. The purpose of the general penalty system is to ensure that TSCA civil penalties are assessed in a fair, uniform, and consistent manner; that the penalties are appropriate for the violation; that economic incentives to violate TSCA are eliminated; and that people will be deterred from committing TSCA violations.

To calculate a fine, EPA uses a two-step approach. First, the agency determines the "gravity-based penalty." This evaluation is based on the nature of the violation, the extent of environmental harm that could result, and the circumstances of the violation. Second, EPA will either decrease or increase the penalty based on the firm's culpability, history of violations, ability to pay, ability to continue in business, and such other matters as justice may require (e.g., self-disclosure of the violation).

The agency organized these factors into a matrix. EPA then tailored the penalty matrix for three TSCA programs—PCBs, inventory reporting, and premanufacture notification. The PCB penalty policy, first issued in 1980, was revised in 1990.

EPA also can reduce a fine if a company agrees to an environmentally beneficial expenditure. Such an expenditure cannot be for some action that is required by law. In addition, the expenditure must exceed the penalty reduction. These expenditures, unlike TSCA penalties, are tax-deductible business expenses, are often referred to as pollution prevention initiatives.

Examples

The following are examples of enforcement actions:

- *An Administrative Proceeding:* The Monsanto Co. agreed to pay a reduced penalty and conduct an audit of all of its health effects studies to settle charges that it failed to report significant risk information as required by TSCA Section 8(e). As part of the audit, the company agreed to pay preset fines for studies identified by the audit that should have been sent to EPA. The company paid a reduced fine of $198,000 for allegedly being five years late in reporting the findings of a cancer study, plus $648,000 in stipulated penalties (*In re* Monsanto Co., TSCA 89-H-21, agreement approved Jan. 3, 1990).
- *A Federal Court Case:* An agreement to settle charges that a pipeline company illegally disposed of PCBs along 10,000 miles of its pipeline brought EPA one of the largest penalties in the agency's history, $15 million. PCBs were used as lubricating oils and fire retardants in 24 turbine compressors at 19 pipeline compressor stations. The company is expected to spend at least $400 million to clean up the contamination and to pay the possible additional fines from lawsuits brought by states with contaminated

sites. The extensive contamination also prompted several congressional hearings (*U.S. v. Texas Eastern Transmission Corp., d/b/a Texas Eastern Gas Pipeline Co.*, 923 F.2d, 32 ER Cases 1822 (5th Cir. 1989)).

- *A Criminal Case:* Two of six people who pleaded guilty to conspiracy to defraud EPA were sentenced to two years in prison. The defendants, company officials from the defunct Martha C. Rose Inc., were charged March 13, 1989, on a 43-count indictment. The charges included conspiracy, false statements, mail fraud, and violations of TSCA for improper PCB storage and disposal between 1982 and 1986. About 13.5 million pounds of PCB-contaminated materials were abandoned by the firm (*U.S. v. Carolawn*, No. 89-0045-01/04-CR-W-JWO (W.D. Mo. Apr. 5, 1990)).

II. FEDERAL INSECTICIDE, FUNGICIDE, AND RODENTICIDE ACT

OVERVIEW

The first U.S. pesticide law was enacted in 1910 to protect consumers from false or misleading claims. The Insecticide Act was administered by the Department of Agriculture, which determined whether a product's label and content matched. Concern over labeling continues to play a significant role in pesticide regulation.

The Federal Insecticide, Fungicide, and Rodenticide Act (FIFRA) was enacted in 1947. Although FIFRA was still primarily a labeling law that prohibited the sale of "adulterated" pesticides, it also required companies to register their products with the federal government before they could be marketed or used in the United States. Subsequent amendments—in 1964, 1972, 1975, 1978, 1980, and 1988—focused on refining the pesticide registration process.

In the 1970s, administration of FIFRA shifted to the newly created Environmental Protection Agency (EPA).

Public concern about adverse effects from pesticides prompted Congress to revise the law and require companies to send EPA safety and product performance data (efficacy data) as part of the pesticide licensing process. The data are evaluated to identify potential risks associated with using the pesticide.

Pesticides are broadly defined under FIFRA Section 2. They include products with an "active ingredient" that is intended to kill, repel, or control unwanted nonhuman life; disinfectants used to kill bacteria in homes and hospitals; herbicides that are used to stop plant growth or defoliate; and traps for certain pests. Along with chemical pesticides, EPA regulates products that control plant growth, including biological products such as pheromones (sex attractants).

Throughout the 1980s, Congress sought a way to provide uniform registration requirements for old and new pesticides. "Old" pesticides are those that were approved before the new registration requirements took effect in 1984. To streamline review of the older pesticides, Congress authorized EPA to group pesticides by active ingredients and develop registration standards for each group.

In the 1988 amendments to FIFRA, Congress made reevaluation of older pesticides a priority with a five-phase program to review all pesticides registered before November 1, 1984. This "reregistration" process is scheduled to be completed by 1997.

In addition to developing pesticide registration (Section 3) and reregistration (Section 4) programs, FIFRA authorizes EPA to regulate other aspects of pesticide use. Under the law, EPA can:

- Issue experimental use permits (EUPs) to firms that need to develop data to meet registration requirements (Section 5)
- Suspend, cancel, or restrict all or part of a pesticide's registration (Section 6) if EPA concludes the product presents an unreasonable risk
- Require firms to keep records and report annual production and use information (Section 7)

- Issue emergency use permits to state or federal agencies that allow use of an unregistered pesticide in an emergency (Section 18)
- Require companies holding canceled or suspended registrations to recall, transport, store, and/or dispose of the affected products (Section 19)

As part of the registration process, EPA evaluates risks associated with pesticide residues on raw agricultural commodities. The agency sets a tolerance level—essentially a limit on the amount of pesticide residues—for certain agricultural products. Congress authorized regulation of these residues under Section 408 of the Federal Food, Drug, and Cosmetic Act (FFDCA). EPA shares responsibility for evaluating residues with the Food and Drug Administration (FDA) and the Department of Agriculture. EPA also sets pesticide limits under FFDCA Section 409, which covers food additives.

U.S. restrictions on certain pesticides have been a contentious international trade issue. Section 17 of FIFRA imposes import and export restrictions on pesticides. FIFRA Section 17(a) requires U.S. companies to meet labeling and notification requirements when exporting unregistered products. Imported products must meet the requirements of FIFRA.

Although FIFRA preempts most state regulation, it does require some state and federal cooperation. States run certified applicator programs, issue experimental use permits, and are involved in registration of pesticides to meet special local needs under Section 24(c). Congress also gave states primary enforcement responsibility. Sections 26 and 27 of FIFRA outline the criteria for states to obtain enforcement responsibility and the conditions under which EPA can revoke state authority.

Violations are often detected through inspections by state authorities (Section 12). If a violation is discovered, EPA can issue a "stop sale, use or removal order" to prohibit distribution of a product (Section 13). The federal government also has the authority to seize or to obtain a court-ordered injunction to stop distribution of an illegal product.

Section 14 allows EPA to assess civil penalties of up to $5,000 for each violation committed by a registrant, commercial applicator, wholesaler, dealer, retailer, or other distributor. Criminal fines may also be imposed when someone knowingly violates the law. Criminal violations of FIFRA are misdemeanors, punishable by no more than one year in jail and fines up to $25,000.

PRODUCT REGISTRATION

A pesticide must be registered if it is any substance or mixture that is intended to be used to prevent, destroy, repel, or mitigate a pest. The definition of pesticide also includes products used to dry out, defoliate, or regulate the growth of a plant. Section 3 of FIFRA requires EPA to register pesticides and license uses of each product so that the pesticide will not cause "unreasonable adverse effects on the environment" if used in accordance with its labeling instructions. The term "environment" under FIFRA includes water, air, land, plants, humans, animals, and any interrelations among these. Registration requirements are codified at 40 CFR 152.

Firms that intend to make, formulate, import, or distribute a pesticide product in the United States must apply to EPA for registration. Distributors and sellers must register a substance if they claim the product contains an "active ingredient" and should be used as a pesticide or could be used to make a pesticide. Other chemicals used to formulate the product are "inert ingredients."

The registration requirement applies to new combinations and mixtures of already registered pesticides as well as to newly discovered chemicals. A company must complete the registration process even if the product is identical in composition to an existing, approved pesticide (40 CFR 152.15).

Although devices intended to trap, destroy, repel, or mitigate a pest are not subject to FIFRA Section 3, other requirements of FIFRA apply. These include branding and labeling requirements, reporting requirements under Section 7, inspection requirements under Section 9, import and export rules under Section 17, and child-resistant packaging under Section 25(c)(3).

EPA publishes formal notices of pesticide registration policies and any changes or clarifications to existing policies. For example, in 1986 the agency issued instructions on how to assemble a package of data for the agency's review (PR Notice 86-5). Noncompliance with a notice can result in suspension of a product's registration.

Registration Procedures

The registration form (EPA Form 8570-1) must be completed by anyone seeking to register a new product or amend an existing registration. EPA will provide forms upon request. The completed form and required data must be sent to EPA headquarters in Washington. EPA publishes a notice of receipt of each application in the *Federal Register*.

In the 1988 amendments, Congress set out a timetable for EPA to respond to new registration applications. EPA must notify the registrant within 45 days of receiving the application whether it is complete, and the agency has 90 days after receiving a complete application to grant or deny the request (Section 3(c)(3)).

If an application is incomplete, EPA notifies the registrant and allows it 75 days to correct deficiencies. In 1989, the agency issued a notice explaining its procedures for expedited reviews (PR Notice 89-2).

In addition to the form, a complete registration must include:

- A list of the data submitted with the application

- A Statement of Formula form, including the name and percentage by weight of each active and inert ingredient (EPA Form 8570-4)
- Five legible copies of draft labels
- Three copies of the required data (40 CFR 158) and, if needed, written permission from the original data submitter granting authorization to use the information
- A statement certifying that the product meets EPA's criteria for child-resistant packaging (40 CFR 162)
- A proposed classification regarding whether the product should be approved for general use or for restricted use only by certified applicators (see the Classification section of this chapter)
- A statement on tolerances if the product's intended use could result in pesticide residue on food or feed crops. If a maximum legal limit for the pesticide's residue has not been authorized under Section 408 or 409 of the FFDCA, the application must be accompanied by an FFDCA petition and appropriate fee to request the establishment of a tolerance or an exemption (40 CFR 180) (see the discussion later in this chapter of pesticide residue/tolerances and the FFDCA)

Pesticide distributors and sellers must send EPA a completed Notice of Supplemental Registration of Distributor form (EPA Form 8570-5). This form must be submitted if a product contains an active ingredient and could be used as a pesticide or to make a pesticide.

Exemptions

FIFRA provides only a few exemptions from registration. Pesticides are exempt if they are regulated by another federal agency; used for exempted purposes, such as protecting wood products or embalming; or distributed, sold, or transferred under certain limited circumstances, such as transfers solely for export (40 CFR 152, Subpart B).

Exclusions

There are a few categories of products that do not have to be registered under FIFRA and certain types of modifications to registrations that do not require an amended registration application.

Products that are not intended for use against pests, such as fertilizer products (those without pesticides) and others intended to aid the growth of desirable plants, are not required to be registered. Another group of products is composed of bleaches, deodorizers, and cleaning agents, unless a pesticidal claim is made on their label (40 CFR 152.8, 152.10).

A company may modify its registration in certain ways without having to amend its application, as long as EPA is notified before the modified product is distributed or sold. For example, EPA needs to be notified only of the addition or substitution of a brand name to a registration. A company is permitted to make corrections to typographical or printing errors on a label, change the package size, and revise certain parts of a label's format without notifying or seeking EPA's approval (40 CFR 152.46).

Data Requirements

When registering a pesticide, EPA weighs the risks and benefits associated with the product. If a pesticide's economic, social, and environmental benefits outweigh the associated risk, EPA will register the product. Companies that apply for new and amended pesticide registrations, reregistration, and experimental use permits are required to provide EPA with data on their product (40 CFR 158).

Because data represent an essential part of obtaining and keeping pesticide registrations, the amount of information required and how EPA uses it have been contentious issues. Data also is a valuable resource. A firm can spend hundreds of thousands of dollars to develop data required by FIFRA.

Product Chemistry

Applications for new and amended registrations and reregistration must include product chemistry data if such data have not been previously reported to EPA. The agency has codified data requirements to support such applications at 40 CFR 158, Subpart C.

Applicants must provide product composition data, a description of the materials used to produce the product, a description of the production and formulation process, a discussion of the formation of impurities, and a description of the product's chemical and physical characteristics.

Coloration

Among the physical characteristics that must be reported is the product's color, which is used as an identifier for safety purposes. EPA can require that pesticides be colored or discolored if necessary to protect health and the environment under FIFRA Section 25(c)(5). For example, several arsenic-based products, including lead arsenate, must meet color standards. The agency has specified that the Munsell Manual of Color or its equivalent be used as a color standard.

Data Requirement Tables

Toxicity, safety, and environmental effects data are required to accompany registration applications. EPA has codified the data requirements in 11 tables. The tables contain the type and minimum amount of data EPA requires to decide whether to approve applications (40 CFR 158, Subpart D).

Some studies in the tables must be completed in all cases, while others are required at EPA's discretion. The type of data required depends on the product's proposed use, results of other tests, and other factors. EPA also can require data beyond the specifications in the tables. The data generally fall into eight categories:

- Residue chemistry data, used by EPA to estimate the exposure of the general population to pesticide residues in food and for setting and enforcing pesticide residue limits in food and feed.

- Environmental fate data, used to assess the toxicity to humans from exposure to pesticide residues remaining after application. Pesticide degradation, metabolism, mobility, dissipation, and accumulation studies often are conducted.

- Data on hazards to humans and domestic animals, assessed through a variety of short- and long-term toxicology studies. Tests to assess mutagenicity, reproductive effects, metabolism, and the cancer-causing potential of certain products are used.

- Reentry protection data, to assess the hazard to farm employees of reentry into areas treated with pesticides and to develop reentry standards for these workers.

- Pesticide spray drift evaluation data, used to estimate exposure to the product. The data help determine the need and appropriate wording for precautionary labeling to minimize the potential adverse effect to nontarget organisms.

- Data on hazards to nontarget organisms, from tests on animals and plants. These wildlife and aquatic organism studies are arranged in a tiered system that starts in the laboratory and ends with field tests.

- Data on product performance (efficacy), to ensure that pesticide products will control the pests listed on their labels.

- Data on biochemical (e.g., pheromones) and microbial (e.g., bacteria, fungi) pesticides, including product analysis, residue analysis, effects on nontarget organisms, and environmental fate of the product.

Companies are required to "flag" certain types of information when sending it to EPA. For example, results from a neurotoxicity study in which the treated animals show a response indicative of acute delayed neurotoxicity, as compared with the control animals, must be flagged.

Minor-Use Pesticides

FIFRA Section 3(c)(2)(A) allows EPA to adopt flexible data requirements for minor-use registrations and tolerance requests. A minor-use pesticide is one that is used on crops planted on a small amount of acreage.

Congress directed EPA to develop data requirements for minor-use registrations. The lawmakers concluded there was insufficient economic incentive for pesticide makers to develop data needed to register a new use under FIFRA's registration provisions or the FFDCA's pesticide tolerance provisions.

In 1986, EPA issued a policy statement on the minor uses of pesticides that outlined an alternative approach to setting tolerances. For example, EPA will consider proposals for tolerance proposals based on pesticide residue data from a limited geographical area for crops with low dietary intake—less than 100 million pounds produced per year, less one pound consumed per person per year, and lack of animal feed concerns. The agency listed more than 200 crops that will be considered automatically for a pesticide tolerance based on representative data. The list includes asparagus, basil, coconuts, dates, garlic, leeks, okra, raspberries, swiss chard, and watercress.

Data Call-In

Under FIFRA Section 3(c)(2)(B), EPA can require pesticide registrants to file additional data to maintain an existing registration—which EPA carries out in its Data Call-In (DCI) program. Within 90 days of receipt of a DCI notice, all registrants of the pesticide must either send the requested data to EPA or provide EPA with evidence that they are taking appropriate steps to secure the data. In addition, they can promise to develop the data either alone or with other registrants.

Companies can also:

- Use the "formulators' exemption" and explain that the firm is exempt from the data requirement because the company buys a registered pesticide for use in formulating another product. FIFRA Section 3(c)(2)(D) exempts formulators from developing safety data for such ingredients; however, the company is not exempt from developing efficacy data documenting the product's performance (benefits). A confidential Statement of Formula (EPA Form 8570-4) must be submitted with the agency to support the claim.
- Request a minor-use exemption under FIFRA Section 3(c)(2)(A) because total production is of small quantity.
- Request a voluntary cancellation or ask EPA to issue a notice of intent to suspend. (See pages 91-92 for a discussion of registration cancellation and suspension.)

EPA also has a policy for companies that need more time to respond to a DCI notice (PR Notice 85-5).

Pesticide Assessment Guidelines

EPA has issued 12 Pesticide Assessment Guidelines. They are listed in 40 CFR 158 along with the pesticide data requirement tables. Each table entry contains a reference number that identifies a specific guideline.

The guidelines, usually less than 100 pages in length, include standards for conducting tests, guidance on evaluation and reporting of data, and examples of test protocols. These documents serve the same function as test guidelines issued under Section 4 of the Toxic Substances Control Act (TSCA), which regulates testing of industrial chemicals. Pesticide assessment guidelines were issued on:

- Product chemistry
- Hazard evaluation for wildlife and aquatic organisms
- Hazard evaluation for humans and domestic animals

- Product performance
- Experimental use permits
- Hazard evaluation for nontarget plants
- Reentry protection
- Hazard evaluation for nontarget insects
- Biochemical pesticides
- Environmental fate
- Residue chemistry
- Spray draft evaluation

The pesticide assessment guidelines, not enforceable by EPA, can be purchased from:

The National Technical Information Service, 5285 Port Royal Rd., Springfield, VA 22161; (703) 487-4650.

Good Laboratory Practices

EPA established standards for good laboratory practices (GLPs) for studies that are required by FIFRA. The GLPs are similar to those issued under TSCA and those issued by the Food and Drug Administration. EPA issued the pesticide GLPs in 1983 and revised them in 1989 (40 CFR 160).

Firms sending EPA test data to fulfill application requirements must include a signed statement to certify the study was conducted according to good laboratory practices. A GLP certification must accompany:

- An application for registration, amended registration, or reregistration under FIFRA Sections 3, 4, or 24(c)
- A submission of data in response to a data call-in under FIFRA Section 3(c)(2)(B)
- An application for experimental use permit under FIFRA Section 5

- An application for an emergency exemption under FIFRA Section 18
- A petition for a food tolerance or food additive regulation under Section 408 or 409 of the FFDCA.

FIFRA laboratory standards cover a tester's organization, personnel, equipment, and facility operations. The testing facility must maintain all raw data, documents, records, protocols, correspondence, and any material germane to interpretation and evaluation of the study. Records may be originals, photocopies, microfiche, or microfilm.

When the study is used to support a research or marketing permit application, the materials must be stored for a minimum of five years. If a test facility closes, all stored records must be transferred to the test sponsor. The sponsor must notify EPA of such a transaction.

Meeting Data Requirements

Companies can satisfy data requirements for pesticide registration, reregistration, or amendments to a registration by conducting the required studies or citing data generated by other companies. However, the applicant may have to pay to use another company's data.

Data submitters' rights are protected by regulations that outline two methods for citing another company's data, define exclusive-use-of-data situations, and spell out procedures for original-data submitters to challenge companies that violate the rules.

Four types of pesticide applications are exempt from the regulation, although they still must meet certain data requirements: applications sent to states under FIFRA Section 24(c), applications for experimental use permits under FIFRA Section 5, applications for emergency exemptions under FIFRA Section 18, and applications from formulators (40 CFR 152, Subpart E).

Citation of Data

Applicants have two options for citing existing studies: the cite-all method (40 CFR 152.86) or the selective method (40 CFR 152.90).

Under the cite-all method, applicants must cite all relevant data sent to EPA by other companies. An applicant can establish a right to cite such data either by obtaining permission or offering to pay each company. If the data are subject to exclusive-use protection under FIFRA Section 3(c)(1)(D), permission must be obtained from the original submitter. The applicant must certify on a data submitter's list, also known as the Data Reference Sheet (EPA Form 8570-20), that the original data submitter authorized use. This form is one of the items required to complete a pesticide registration application.

The selective method provides more alternatives, but an applicant still must meet the data requirements included in a registration standard or the data requirement tables. Compliance with the requirements can be demonstrated by:

- Obtaining a waiver of the data requirement (40 CFR 152.91)
- Citing the applicant's own new study or previously submitted study (40 CFR 152.92)
- Citing the public literature (40 CFR 152.94)
- Citing previous data submitters or exclusive-use data with the owner's permission (40 CFR 152.95)
- Documenting that no previous data (a so-called data gap) have been submitted to EPA

Data-Use Compensation

The requirement to compensate for data use was included in FIFRA Section 3(c)(1)(D) in the 1970s, when data use was limited by whether the information represented a trade secret. The data compensation controversy spawned many lawsuits, which brought the registration program to a standstill in the mid-1980s.

The need to resolve this controversy and register "me-too" products (those that are substantially similar to existing registrations) led to a 1984 U.S. Supreme Court decision upholding EPA's disclosure of previously submitted data in the interest of registering "me-too" products (*EPA v. Monsanto Co.*, 467 U.S. 986, 21 ER Cases 1062 (1984)).

When compensation is required, the amount is fixed by arbitrators appointed by the Federal Mediation and Conciliation Service (FMCS). Cost-allocation disputes that stem from joint testing efforts also are settled by binding arbitration. FMCS's regulations for arbitration of pesticide data disputes are codified at 29 CFR 1440.

In 1985, the Supreme Court again addressed the constitutionality of FIFRA Section 3(c)(1)(D) when several pesticide manufacturers argued that the FIFRA arbitration scheme delegated too much power to arbitrators without sufficient guidance from Congress. The high court upheld the delegation of power. (*EPA v. Union Carbide Agricultural Products Co.*, 105 U.S. 3325, 21 ER Cases 1984 (1985)).

Formulator's Exemption

An applicant who wants to register a new pesticide use and who buys a registered pesticide from another producer is not required to send or cite any safety data on the purchased product under FIFRA Section 3(c)(2)(D). However, there is still a reporting obligation (40 CFR 158.50).

The applicant must complete a Formulator Exemption Statement, identifying the active ingredient for which an exemption is sought. The applicant also must file a confidential Statement of Formula (EPA Form 8570-4) with the agency. Once this information is on file, the firm needs to resubmit it only when the company changes its source for an active ingredient.

Exclusive Use

Data owners who registered pesticides after enactment of the 1978 FIFRA amendments are entitled to 10 years of exclusive use of data sent to EPA in support of that application (Section 3(c)(1)(D)(i)). The original data submitter may give written permission to other parties to use the data during that time period.

Two types of data are eligible for this classification: data used for the original registration of the product containing a new active ingredient and data submitted to support an amendment adding a new use to the registration that pertains only to the new use.

"Defensive data," data reported in response to a data call-in notice or a special review, are not eligible for such protection, nor are data sent to EPA before 1978.

Once the exclusive-use period expires, EPA may consider the data in support of an application by another registrant without the permission of the original data submitter.

Trade Secrets

There are agencywide and FIFRA-specific rules to protect a registrant's trade secrets. FIFRA Section 10 provides that test data may not be disclosed until a registration or pesticide tolerance is issued. The statute also prohibits the disclosure of test data to individuals affiliated with international commerce in pesticides.

Financial data also are protected. Chemical formulas of products, however, even if claimed as confidential, may be revealed by EPA in certain situations. For example, product chemistry data may be released during a registration cancellation hearing. EPA's general rules for handling confidential information include FIFRA-specific provisions (40 CFR 2.307).

Public Access

Information sent to EPA in support of a registration becomes part of the agency's file on the registration. Within 30 days after registration, EPA makes the data available for public inspection. Materials that will be publicly available include the applicant's list of data requirements, the method used by the applicant to demonstrate compliance with each data requirement, and the applicant's citations of specific studies in the agency's possession, if applicable (40 CFR 152.119).

The data on which EPA based its registration decision will be made available for public information, except for data protected by FIFRA Section 10, protection of trade secrets (40 CFR 2).

Fees

In 1988, FIFRA was amended to include two types of fees to support EPA's pesticide registration program: maintenance fees and reregistration fees. Congress mandated the fee scheme in FIFRA Section 4(i). Fees will continue to draw congressional attention during the 1990s as EPA tries to meet its mandated timetable for reregistration.

Reregistration fees must be paid in phase two (when a company commits to generating data) and in phase three (when the data are sent to EPA) of the five-step reregistration process. Companies divide the two reregistration fees based on their market share of the product. The fees range from $50,000 to $100,000. Exemptions from the fees are allowed for minor uses and some antimicrobial products.

A maintenance fee of about $650 for each product registration must be paid on March 1 of each year. EPA is scheduled to collect these fees for nine years, from 1989 to 1997. The exact amount a company pays depends on the number of registrations it holds. Congress has also established ceilings for the total amount companies pay.

CLASSIFICATION

Section 3(d) of FIFRA requires EPA to classify a pesticide product and its use. Pesticides can be designated for restricted use or general use. Some older products are unclassified because they were registered before the classification system was developed. The classificaton is based on characteristics of the product or its use—the product's toxicity, flammability, and effect on nontarget organisms are some of the criteria (40 CFR 152, Subpart I).

EPA has authority to impose restrictions on the product's composition, labeling, packaging, uses, distribution and sale and on the qualifications of the user. As of 1988, EPA had categorized about 50 pesticide products for restricted use.

Classification reviews generally occur as part of EPA's review of the application for registration of a new product. The agency can conduct classification reviews at any time, however, if it determines that a restriction on the use of a product is necessary to avoid unreasonable adverse effects on the environment.

Restricted-use products may only be used by, or under the direct supervision of, a certified applicator. FIFRA mandates that state or federal programs be established and approved by EPA to train and certify applicators. The Department of Agriculture's Agricultural Cooperative Extension Service usually helps train applicators. (See the next section of this chapter for a discussion of state programs to certify applicators.)

The distribution and sale of restricted-use pesticides is also regulated by EPA. For example, any advertisements, labels, or brochures must clearly state that the product is a "restricted-use pesticide."

Products with a general-use classification may be sold with no restrictions on who may use the pesticide. In general, the criteria are more stringent for a general-use classification than for other classifications.

Older pesticides that were registered before the classification system was established are now scheduled to be reregistered. At that time, the substances will be classified.

APPLICATOR CERTIFICATION

Application of restricted-use pesticides must be performed or supervised by a certified applicator under FIFRA

Section 3(d)(1)(C). There are two main classifications of certified applicators—private and commercial (Section 11). In 1991, there were about 1 million private and 300,000 commercial certified applicators, according to EPA.

EPA has developed standards for certification and criteria for state-administered training programs. Most commercial applicator's permits are valid for three years and private applicator's permits for four years. Generally, renewal and recertification require applicators to take a written examination. Commercial applicators also must maintain records for two years following each application (40 CFR 171).

States, with assistance from EPA, generally train and certify applicators. States may administer their own applicator certification program if it meets EPA standards. In states with no certification program, such as Nebraska and Colorado, people who want to be certified must contact EPA. The agency also provides for certification of applicators on Indian reservations.

Commercial Applicator Standards

A commercial applicator is a certified applicator who uses or supervises the use of restricted-use products on another party's property for a fee. There are 10 categories of commercial applicators:

- Agricultural pest control
- Forest pest control
- Ornamental and turf pest control
- Seed treatment
- Aquatic pest control (weed control)
- Right-of-way pest control (e.g., roadsides, pipelines, tank farms, and fence lines)
- Industrial, institutional, structural, and health-related pest control (e.g., hospitals)

- Public health pest control
- Regulatory pest control
- Demonstration and research pest control

In 1991, approximately 98,500 commercial applicators were registered for ornamental and turf pest control; approximately 87,200 were registered for industrial, institutional, structural, and health-related pest control; approximately 40,000 were registered for right-of-way pest control; and approximately 12,000 were registered for aquatic pest control.

Written and performance tests are used to determine the competence of commercial applicators. Prospective applicators must meet both general and specific standards for each category of commercial pesticide application. All commercial applicators must demonstrate practical knowledge in areas such as label comprehension, safety, potential environmental consequences of use, pests and pesticides, equipment, and applicable requirements.

Private Applicator Standards

Private applicators must meet many of the same requirements as commercial applicators, but regulators cannot require them to pass a test as a condition of certification. Private applicators may only apply pesticides on land they own or rent for production of agricultural commodities, or for an employer who owns or rents the property. In some cases, a private applicator may apply pesticides on another person's land if that work is performed free of charge as an in-kind service.

A private applicator must demonstrate a practical knowledge of pest problems associated with the applicator's agricultural operations. He or she also must be familiar with proper storage, use, handling, and disposal of pesticides and containers and related legal responsibilities.

The applicator also must be able to:

- Recognize common pests
- Read and understand the label and labeling information
- Apply pesticides in accordance with label instructions and warnings
- Recognize local environmental conditions that could lead to contamination
- Recognize poisoning symptoms and the procedures to follow in case of an accident

States may require private applicators to take training courses but cannot require examinations for certification.

PRODUCT LABELING, PACKAGING

Information that will be included on the product's label must be sent to EPA as part of the registration process. FIFRA Section 25(c)(3) authorizes EPA to establish labeling requirements, which are codified as labeling rules at 40 CFR 156.

A pesticide is considered misbranded if its labeling is false or misleading in any way. For example, a false or misleading statement about the product's effectiveness or composition would be considered misbranding. Safety claims made without the qualifying phrase "when used as directed" also are considered misbranding.

Pesticide labels must be written in English and printed on contrasting background. EPA rules specify what information must appear on the front panel of a label. A label must include the following information and be set in six-point type or larger:

- Name brand, or trademark under which the product is sold
- Name and address of the producer or registrant
- Net contents of the product

- Registration number of the product
- Producing establishment's identification number
- Statement of ingredients
- Warning or precautionary statements
- Directions for use
- Use classification

Warning Statements

A label must warn against hazards to children, human health, or the environment and other physical and chemical hazards. Almost every pesticide label must include on its front panel the statement "keep out of reach of children." EPA has designated which warning statements must appear on the front label and what information may appear elsewhere.

The agency has devised a four-tier human hazard classification scheme with a corresponding "hazard signal." The hazard classification is based on data from toxicity studies on oral, inhalation, or skin exposure routes. Classifications also reflect the pesticide's effect on eyes and skin. Hazard signals such as "danger," "warning," and "caution" are required depending on the level of toxicity observed in studies.

If a product poses hazards to nontarget organisms, EPA may require certain precautionary statements. For example, pesticides intended for outdoor uses other than aquatic uses should bear the statement "Keep out of lakes, ponds, or streams. Do not contaminate water by cleaning equipment or disposal waste."

Child-Resistant Packaging

FIFRA Section 25(c)(3) requires that certain pesticides and devices be distributed in packages that are child resistant. Products that exceed toxicity criteria or are labeled for residential use are required to be in special packaging (40 CFR 157, Subpart B). Exemptions to the packaging requirement include restricted-use pesticides and those products packaged in bulk. On a case-by-case basis, EPA may grant additional exemptions.

Child-resistant packages must meet the effectiveness specifications of the Consumer Product Safety Commission's (CPSC's) poison prevention standards (16 CFR 1700). Testing protocols are set forth in the CPSC rules. The registrant must maintain certain records for as long as the registration requires child-resistant packing (40 CFR 157.36). Information that must be retained includes a description of the package, a copy of the statement certifying that the packaging meets federal standards, test data verifying that each package for the product is child resistant, and records verifying that the package meets compatibility and durability standards.

APPLICATION REVIEW STANDARDS

Historically, EPA has reviewed pesticide applications on a chemical-by-chemical basis. In the 1978 and 1988 amendments to FIFRA, Congress tried to improve the registration review proess by directing EPA to develop registration standards for groups of active ingredients.

In 1978, Congress directed the agency to develop registration standards for each of the 600 groups of active ingredients that were used in 35,000 pesticide products. Each registration standard (document) is a comprehensive statement of EPA's position on pesticide products containing the active ingredient. The document also lists the test data, use, and labeling requirements for all products containing the active ingredient.

EPA was slow to develop the registration standards. By 1988, fewer than 50 final standards, known as the Final Registration Standard and Tolerance Review, had been completed. The agency, however, issued about 200 interim registration standards. These documents are available from EPA.

In the 1988 amendments, Congress rewrote FIFRA Section 3—the pesticide registration section—so EPA could devote more resources to reregistration. The agency was directed to decide whether to register a product or use within 90 days of receipt of a complete application or amendment to a registration. The lawmakers also expanded EPA's authority to evaluate applications under either the unconditional registration criteria of FIFRA Section 3(c)(5) or the conditional registration criteria of FIFRA Section 3(c)(7).

This "fast-track" provision was designed to expedite EPA consideration of generic or "me-too" registrations and amendments to registrations. EPA was directed to review as quickly as possible any application that, if registered as proposed, would be "identical or substantially similar in composition and labeling" to a registered product.

EPA determined that it would limit complete data reviews, which are required for unconditional registrations, to applications to register new active ingredients or to special cases. EPA stated, in PR Notice 89-2, that it will use the criteria for conditional registration in FIFRA Section 3(c)(7)(A) and (B) as the basis for all "me-too" application reviews. Review standards are codified in 40 CFR 152, Subpart F.

Conditional Review

The level of review for conditional registration depends on whether the product contains a new active ingredient. This process is also known as registration under special circumstances.

For products that do not contain a new active ingredient, EPA may approve the registration if it determines that it has all the data necessary to evaluate the application, the product would not significantly increase the risk of unreasonable adverse effects on the environment, and the product and its proposed use are substantially similar to currently registered products (40 CFR 152.113).

EPA will not approve a conditional registration of a product or use if the active ingredient is the subject of a special review—a discretionary review of a product or use initiated by EPA based on evidence of unreasonable adverse effects—or if a major food or feed crop is involved.

Each registration issued under this part of FIFRA must include the submission or citation of all data that are required for the unconditional registration of the product under FIFRA Section 3(c)(5).

The agency approves applications for registration of products that contain new active ingredients "for a period of time sufficient for the generation" and reporting of data necessary to complete an unconditional review under Section 3(c)(5). EPA must also determine that the conditional registration will not cause any unreasonable adverse effect on the environment and that it is in the public interest (40 CFR 152.114).

Unconditional Review

EPA approves applications for unconditional registrations (FIFRA Section 3(c)(5)) if it determines that:

- The application is complete
- All relevant data have been reviewed
- No additional data are necessary
- The composition of the product warrants the proposed efficacy claims (FIFRA Section 24(c)) and the required data have been reported
- The product will perform its intended function without unreasonable adverse effects on the environment
- The product is not misbranded and its packaging and labeling requirements comply with FIFRA
- The necessary tolerance or other regulation has been issued under Section 408 or 409 of the Federal Food, Drug,

and Cosmetic Act if the product is to be used on food, animal feed, or feed crops or is reasonably expected to result in pesticide residues

- The requirements of the FDA have been satisfied, if the pesticide product also is a drug as defined by FFDCA

Approval or Denial of Registration

EPA will register a pesticide if the product:

- Warrants the company's claims for it
- Meets the labeling requirements of FIFRA
- Will perform its intended function without unreasonable adverse effects on the environment
- Will not cause unreasonable adverse effects on the environment when used following widespread and commonly recognized practices

EPA publicly announces all registrations. The notices include a description of the new chemical or new use and summarize the agency's regulatory decisions, cite missing data, and respond to comments.

EPA may deny an application for registration if it determines the product does not meet the criteria under FIFRA Section 3(c)(5) for unconditional registration or under FIFRA Section 3(c)(7) for conditional registration (40 CFR 152.118).

The agency will notify the applicant by letter of the denial and explain the basis for the determination and the conditions, if any, that must be met for the registration to be approved. EPA also announces the denial in the *Federal Register*. A company has 30 days to correct the deficiencies identified by EPA, withdraw the application, or request a hearing.

EMERGENCY EXEMPTIONS

Congress authorized EPA to issue emergency use exemptions to state and federal agencies under certain emergency conditions (Section 18). This provision was designed to allow government agencies to respond quickly to emergency pest situations. Exemptions can be sought for any provision of the pesticide law. EPA has established four types of exemptions and a general application process (40 CFR 166).

There are four types of emergency-use exemptions.

- Specific exemptions, which permit the use of an unregistered product to avert a significant economic loss or a significant risk to endangered species, threatened species, beneficial organisms, or the environment. Specific exemptions are granted only for as long as expected to be necessary, and in no case for more than one year.
- Quarantine exemptions, to control the introduction or spread of any pest new to or not known to be widely prevalent in the United States. The exemption may last only for as long as necessary, and in no case more than three years.
- Public health exemptions, to control pests that would cause a significant risk to human health. These exemptions are granted only for as long as necessary, and in no case for more than one year.
- Crisis exemptions, used when there is insufficient time to request a specific, quarantine, or public health exemption or if such a request has been made but there is insufficient time for EPA to complete the review. Crisis exemptions may be issued only for 15 days, unless an application requesting one of the other emergency use exemptions has been filed with EPA.

An emergency condition is defined as an urgent, nonroutine situation that requires the use of an unregistered product when no effective pesticide is registered for the needed use; no economically or environmentally feasible alternative exists; and the situation involves a new pest, a significant risk to human health, or significant economic loss.

Applications for specific, quarantine, or public health exemptions must be filed by the head of the federal or state agency involved or by a governor or other designated official.

The exemption request must describe the emergency, the pesticide proposed for use, and the nature of the proposed use. A detailed explanation of why existing, registered pesticides cannot be used also is required. The applicant must provide EPA with data to prove that the proposed use will be effective to control the emergency. A list of foods likely to contain residues and a discussion of risk information must be included in the application.

EPA publishes a notice of each exemption request and invites public comment for 15 days. The agency also announces the approval of any exemption. If residues will result from use, EPA also will contact FDA, the Department of Agriculture, and state health officials. Under certain circumstances, EPA may revoke an exemption.

These Section 18 exemptions also carry reporting and recordkeeping requirements. Any unexpected adverse effect must be reported to EPA immediately. A report summarizing the results of the pesticide use under a specific, quarantine, or public health exemption must be sent to EPA within six months after the exemption expires. Records must be maintained for at least two years whenever a use involved a food crop.

The head of a state or federal agency or a governor may issue crisis exemptions, but only during emergencies and only when there is insufficient time to request one of the other exemptions.

Crisis exemptions may not be used to authorize use of suspended or canceled pesticides, new pesticides, or a new food use of a pesticide. The authority that issued the crisis exemption must notify EPA within 24 hours.

When a crisis exemption is about to be or already has been issued, EPA immediately undertakes an expedited review of the pesticide. The agency has authority to revoke a crisis exemption at any time. EPA will notify other relevant food and health agencies as appropriate. EPA also will publish periodic notices and an annual report summarizing the number of crisis exemptions declared and the number of revoked exemptions. The reporting and recordkeeping requirements for crisis exemptions are comparable to other types of FIFRA Section 18 exemptions.

GENERAL REPORTING, RECORDKEEPING

FIFRA Section 8 allows EPA to require companies to maintain records of their production, EPA also has authority to inspect the records of any distributor, carrier, dealer, or company that offers for sale, delivers, or offers for delivery any pesticide. Records of "delivery, movement, or holding of a pesticide or device, including the quantity, the date of shipment and receipt, and the name of the consignor and consignee" must be maintained (40 CFR 169).

Establishment Registration

All facilities that produce pesticides and pesticidal devices must be registered with EPA and file annual reports. FIFRA Section 7 requires companies to register their establishment or place of business. EPA reported that it received 7,836 pesticide production reports for 1989. This registration requirement also applies to firms that produce a pesticide or active ingredient for export as well as foreign and U.S. companies that use a pesticide under a FIFRA Section 5 experimental use permit, a FIFRA Section 18 emergency exemption, or a FIFRA Section 24(c) special local needs permit. The registration requirement does not apply to custom blenders (40 CFR 167).

Registration of the establishment must be completed before any pesticide production begins. To register, a firm must send EPA the name and address of the company, the type of ownership, and the names and addressses of each facility for which registration is sought.

On approval of a registration, the agency assigns the establishment a number that must be used on every pesticide product's label or container. Changes in ownership or address must be sent to EPA within 30 days of such changes.

Annual Report

Even if the registered establishment does not produce pesticide products during a reporting year, the establishment must file an annual report. If a pesticide maker did not produce any pesticides, the annual report must state "zero production."

The report (EPA Form 3540-16) must include information about the amount of the product produced in the past year, the amount sold or distributed in the past year, and estimated current-year production. EPA monitors this reporting closely, and sues firms that fail to file their reports.

In addition to the annual report, companies that produce pesticides must file a chemical formulation report (EPA Form 3540-16) describing pesticide production, sales, and distribution in the past year. The chemical formulation report is treated as confidential business information under Section 7(d) of FIFRA.

Producers' Recordkeeping

Pesticide producers must keep records on production, rcccipt of shipments, shipping, disposal, and testing for two years. Producers also must keep inventory records on the amounts and types of pesticides in stock until a more current inventory is prepared. Pesticide producers also must keep records of:

- All domestic advertising for restricted-use pesticides, including scripts from radio and television advertising, for two years

- The method, dates, and locations of pesticides disposal, as well as the type and amount of the product, for 20 years or maintain them for three years and then forward them to EPA
- Tests conducted on human beings for 20 years or maintain them for three years and forward them to EPA
- Adverse effects on the environment from the pesticide, as long as the registration is valid or the producer is in business
- Research data, as long as the registration is valid and the producer is in business

Recordkeeping for Dealers of Restricted-Use Pesticides

Records of all transactions by a dealer of restricted-use products must be kept for two years. Dealers of restricted-use pesticides can make such products available only to certified applicators. Uncertified applicators may obtain such products only if they can document that the pesticide will be used by a certified applicator (49 CFR 171.11).

Recordkeeping for Experimental Use Permits

Companies holding a FIFRA Section 5 experimental use permit must file a report with EPA every three months. Information about the sale and distribution of the product must be reported, along with a summary of the progress made toward obtaining data during the previous three months. Within six months of the permit's expiration, a final report must be filed with EPA.

Adverse Effects Reporting

FIFRA Section 6(a)(2) requires that companies report "additional factual information regarding the unreasonable adverse effects on the environment" of a pesticide. Such information must be reported within 15 days of the date it was obtained (40 CFR 153, Subpart D).

This reporting requirement applies to any party acting for or employed by the registrant, generally any company that makes, formulates, or distributes a registered pesticide product. The person must be "capable of appreciating the significance of such information." This information should be reported if EPA is unaware of it and it pertains to a pesticide for which the registrant holds a registration under FIFRA.

Companies also must comply with the requirements under FIFRA Section 3(c)(2)(B) and/or Section 3(c)(1)(D) for flagging data demonstrating possible adverse effects. (See the section on Data Requirements in this chapter for a discussion of flagging test data.)

EPA has provided guidance on how to report data from toxicological, epidemiological, efficacy, and residue studies (40 CFR 153.69–153.73). Certain types of information need not be sent to EPA: (1) information that consists solely of the opinion or conclusion of a nonexpert, (2) information previously sent to EPA by the company, (3) any scientific article or publication that has been abstracted, or (4) reports written by EPA or other federal agencies.

Substantial Risk Reporting

Although TSCA excludes pesticides from its coverage, the law has one reporting requirement that affects pesticides makers. Under certain circumstances, substantial risk information must be reported promptly to EPA under TSCA Section 8(e).

TSCA reporting, however, depends on whether the adverse effect information concerns an already-registered pesticide. If the risk information relates to a registered pesticide, companies need to comply only with the adverse effects reporting requirement under FIFRA Section 6(a)(2). If the risk information concerns a substance that is not registered, regardless of whether the pesticide maker intends to file a FIFRA application, the TSCA Section 8(e) reporting requirement applies.

TSCA regulates chemical substances and mixtures. Section 8(e) of TSCA requires anyone who obtains information that a chemical substance or mixture presents an unreasonable risk to human health or the environment to report that information to EPA within 15 days (see the earlier discussion on TSCA reporting requirements).

EPA considers a chemical to be a pesticide when the substance is the subject of an experimental use permit or an application for pesticide registration. Prior to that, EPA treats it as a chemical substance within the jurisdiction of TCSA. Therefore, according to EPA, any substantial risk information on a research and development substance that may arise prior to the experimental use permit or registration applications must be reported under TSCA Section 8(e), even if that information might be sent to EPA as part of a FIFRA application. (See EPA status report 8EHQ-0285-0545 S for more information.)

TOLERANCES, EXPERIMENTAL USE PERMITS

An integral part of EPA's pesticide program is the establishment of tolerances for pesticide residues on raw agricultural products under Section 408 of the Federal Food, Drug, and Cosmetic Act. The agency also has authority under FFDCA Section 409, which regulates food additives, to establish maximum legal limits for certain pesticide residue on processed foods. Foods containing pesticide residues that exceed these limits are considered adulterated under FFDCA. Such foods, if sold or distributed, are subject to seizure and condemnation under FFDCA.

A tolerance must be set before pesticide products can be legally sold, distributed, or used in the United States. EPA sets maximum legal limits for residues, known as tolerances, on an individual chemical basis for specific foods (40 CFR 180).

Other federal agencies also set tolerances. FDA sets pesticide residue limits in processed foods, and the Department of Agriculture sets limits for certain meat products. FDA enforces most tolerances, including those set by EPA. FDA inspects products for pesticide levels, called "action levels." If the raw agricultural product exceeds that level, it is considered adulterated.

Tolerances, expressed in terms of parts per million of the chemical by weight, may be raised, lowered, or revoked. EPA has issued thousands of tolerances.

A company filing a petition to establish a tolerance must also provide EPA with adequate analytical methods and data to determine the total toxic residue, as indicated in FDA's *Pesticide Analytical Manual*. EPA defines this residue to include the parent pesticide and its degradation products, metabolites, and impurities of toxicological concern (PR Notice 88-5).

A temporary tolerance or exemption may be issued with FIFRA Section 5 experimental use permits. These temporary tolerances are needed in cases where the experimental product is likely to result in residue on food or feed crops. The process for obtaining a temporary tolerance is similar to the petition process for a permanent tolerance. Temporary tolerances are not required if an applicant certifies that the experimentally treated crop will be destroyed or fed to animals for testing purposes.

For certain pesticides residues on raw commodities (FFDCA Section 408), EPA has adopted a zero tolerance. This means that no amount of a pesticide chemical may be on the commodity when it is shipped.

Tolerance Petition Process

Manufacturers must petition EPA's registration division to obtain a pesticide tolerance or to amend or repeal a tolerance. The period for review begins once EPA certifies that the pesticide is useful and decides whether the proposed tolerance reasonably reflects the predicted residue. EPA has 90 days after the certification to review the petition.

The petition must include:

- The name, chemical identity, and composition of the pesticide chemical
- The amount, frequency, and time of application of the pesticide
- Full reports regarding the safety of the substance
- Results of tests on the amount of residue remaining and a description of the analytical method used
- Practicable methods for removing any residue that exceeds the proposed tolerance
- Proposed tolerances for the pesticide
- Reasonable grounds in support of the petition

EPA has developed an extensive fee schedule to accommodate all transactions. For example, a fee of $48,225 must accompany a petition for a new tolerance or an increase in an existing tolerance. That fee covers the establishment of tolerances on nine commodities. Each additional commodity requires an additional $1,200.

Either EPA or the petitioner may request that the petition be referred to an advisory committee. The committee will have at least three members qualified to review the safety of the pesticide.

The agency publishes all proposed tolerances in the *Federal Register* and solicits public comment. A final tolerance also is published and codified at 40 CFR 180, Subpart C.

Experimental Use Permits

Because pesticide registrations must be supported by test data, Section 5 of FIFRA was written to enable EPA to issue experimental use permits (EUPs). These permits outline certain conditions under which an unregistered pesticide may be used, applied, transported, and disposed. Most permits are issued for one year but may be renewed (40 CFR 172).

EUPs may be issued to individuals, agricultural research agencies, or educational institutions. EPA may revoke a permit at any time if it finds the permit terms are inadequate to avoid unreasonable adverse effects on the environment. Most states, as well as EPA, can issue EUPs. Permit holders are required to maintain certain records. (See the section on General Reporting, Recordkeeping in this chapter for a discussion of recordkeeping.)

Application Process

The EUP application requires a detailed explanation of intended use, potentially affected crops and animals, information on persons entering the treated area, results of toxicity studies and other data relevant to the product's potential for injuring human health, and other information.

If use of the experimental product could result in residue on food or feed, the applicant faces additional requirements, including obtaining a tolerance or exemption and showing how affected food or animals will be handled.

The agency announces in the *Federal Register* any application or issuance of an EUP. The agency also will provide notice of state-issued permits.

EUPs are not required for certain small-scale testing not designed for economic benefit (40 CFR 172.3); however, there are still restrictions on use conditions. For example, EPA requires that the resulting crops be destroyed.

Applications for EUPs carry a fee of $4,500, which may be shared or waived in certain circumstances.

REREGISTRATION OF OLDER PESTICIDES

When Congress revised FIFRA in the early 1970s, mandating that registration decisions on new pesticides be based on an assessment of the risks and benefits of the product, it created a double standard. Older pesticides that were registered by the Department of Agriculture could be manufactured, imported, sold, and used even though the federal government had not evaluated their safety based on current scientific knowledge.

To eliminate the double standard, Congress directed EPA to review and reregister older pesticides (FIFRA Section 4). Congress also authorized EPA to group pesticides by active ingredients, review available data, and require the submission of additional data instead of reviewing registrations on a product-by-product basis. EPA spent years developing registration standards. (See the section on Application Review Standards in this chapter for a discussion of registration standards.)

Although Congress directed EPA to launch the reregistration process in 1978, it was not until 1988, when FIFRA was again amended, that the agency faced mandatory time frames for completing the process.

Five-Phase Plan

In 1988, Congress set up a five-phase plan to accelerate the reregistration of pesticides registered before November 1, 1984. Registrants that fail to reregister are subject to cancellation. The five phases are to be completed by 1997.

In Phase One, EPA was required to publish lists of all ingredients to be reregistered. List A contained 194 active ingredients for which the agency had issued generic registration standards. Products for which such standards were issued were exempt from phases two through four. Lists B and C each were to contain 150 active ingredients and List D the remaining active ingredients. The lists emphasize food or feed uses and pesticides that might leach into groundwater. These lists were to be published by the end of 1989.

In Phase Two, registrants were required to inform EPA within three months of publication of each list whether they intended to reregister their products. Those seeking to reregister products were required to identify missing or inadequate data. Registrants had to agree to fill the data gaps through cost-sharing arrangements. They also had to pay the first installment, $50,000 of the $150,000 reregistration fee. The agency cancelled the registrations of those that did not seek reregistration. EPA was authorized to allow up to 48 months for the data to be reported. The agency was directed only to grant extensions under "extraordinary circumstances" beyond the registrants' control.

Under Phase Three, registrants must summarize and reformat existing studies, certify the availability of raw data, flag studies that indicate adverse effects, make commitments to satisfy all applicable data requirements, and pay the final installment of the registration fee, $100,000 for each active ingredient. In early 1990, EPA issued a policy on how data summaries should be structured (PR Notice 89-3).

In Phase Four, EPA is required to review the data sent in during phases two and three and direct registrants to fill the gaps or announce that a data base is complete. EPA issued a FIFRA Section 3(c)(2)(B) DCI notice to require registrants to supply the missing data. (See the discussion of Data Call-In notices earlier in this chapter.)

Under Phase Five, EPA reviews the data and determines whether to reregister the product. The schedule for completing this stage ranges from three to nine years after the 1988 amendments.

Application Process

Pesticide reregistration forms should be submitted only when solicited by the agency, and solicitations generally involve groups of similar products.

Before calling for reregistration, EPA determines the sufficiency of data previously sent to the agency, whether the products are candidates for restricted use, and whether the precautionary statements are sufficient. EPA sends out guidance packages to all registrants with licensed products in the group.

When evaluation of the data indicates that long-term studies are needed, EPA may issue a conditional reregistration for the product, which cannot exceed five years.

Congress imposed a reregistration fee of $150,000 per active ingredient for products used on major food or animal feed crops. Reregistration fees for active ingredients not for use on major food or feed crops range between $75,000 and $150,000. Fees are apportioned among registrants based on market share.

Reregistration fees may be waived if an antimicrobial active ingredient is produced in quantities of less than 1 million pounds per year, the active ingredient is considered minor, or the registrant qualifies as a small business. The reregistration fees for small businesses are based on a percentage of average annual sales.

RISKY PESTICIDES

When EPA determines that the risk posed by a pesticide is excessive, it notifies the companies holding registration of its intention to initiate a special review. This is one of the first steps EPA can take to eliminate a risky pesticide. Section 6 of FIFRA authorizes EPA to limit a registration if the substance presents an unreasonable risk to humans or the environment.

EPA also can suspend, cancel, or restrict an existing registration. All final regulatory actions, decisions, or orders under FIFRA Section 6 are subject to review in a U.S. district court (Section 16).

Special Review

The risk management procedure EPA uses to determine whether a pesticide poses unreasonable adverse effects is called the special review process. This process is a discretionary function of EPA. As of 1990, EPA had initiated 59 special reviews and concluded 44 reviews. The process is designed to ensure that EPA assesses the risks and benefits of a pesticide in an open and responsive manner (40 CFR 154).

EPA can initiate a special review based on:

- Adverse effects data submitted under FIFRA Section 6(a)(2)
- Adverse effects data submitted in support of a registration under Section 3(c)(2)(B)
- A generic registration standard
- Data obtained from a federal testing facility, such as the National Cancer Institute, the National Toxicology Program, or the National Institute of Occupational Safety and Health

The agency may decide no change is needed with the registration, cancel some uses, or make modifications to the terms and conditions of the registration at the close of the special review.

During EPA's review of test data, the agency assesses the hazard and exposure associated with each use. The agency looks for the possible short-term and long-term health effects from exposure to the pesticide, as well as other significant risks such as risk to an endangered species.

When EPA decides to initiate a special review, it notifies the registrant and gives the company 30 days to dispute EPA's decision. After reviewing the registrant's arguments, EPA's announces in the *Federal Register* whether it will proceed with the special review and solicits public comment on the decision.

Suspension

Suspension of a pesticide is an emergency procedure that authorizes EPA to suspend a product during cancellation proceedings, which may take years to complete. The agency used this authority during its cancellation proceedings for dinoseb and dibromochloropropane.

The agency may issue a notice of intent to suspend the registration of a product if (1) under FIFRA Section 6(c)(1) the agency determines that suspension is necessary to prevent an imminent hazard during the time necessary for cancellation or change in classification proceedings, or (2) under FIFRA Section 3(c)(2)(B), DCI, the registrant fails to supply the required data. EPA will issue a suspension order that outlines the date of suspension and the terms for disposing of the suspended product. After that date, the distribution, sale, or use of a suspended pesticide is considered a violation of FIFRA.

Cancellation

Cancellation proceedings are adversarial, often involving administrative hearings and court hearings, so EPA seldom resorts to cancellation. EPA begins by notifying the registrants and announcing its intent to cancel a registration in the *Federal Register*. EPA may seek to cancel a registration, change a pesticide's use classification, or hold a hearing to determine what action is needed. A registrant that receives a notice of intent to cancel may request a hearing.

Grounds for cancellation include failure to comply with FIFRA, unreasonable adverse effects on the environment when used in accordance with commonly accepted practices (FIFRA Section 6(b)), and failure to meet any condition of registration.

The cancellation process was used to ban DDT; 2,4,5-T/Silvex; Kepone; mirex; ethylene dibromide (EDB); aldrin/dieldrin; and Compound 1080 when used to control predators.

Voluntary Cancellation

A registrant may request that EPA cancel the registration for a product at any time. If the registrant wants to distribute and sell existing stocks of the product, such a request must be made when the request to cancel is filed (40 CFR 152.138). EPA has persuaded many companies to cancel registrations voluntarily as an alternative to the lengthy special review, suspension, or cancellation processes.

For example, EPA persuaded the makers of a mercury-based biocide to voluntarily cancel their registrations for use as a preservative in indoor latex paint. In December 1989, the agency became aware of an incident where a Michigan child contracted severe mercury poisoning after his home was painted with overformulated paint containing a mercury-based preservative. Although the agency had a lengthy record of evidence and prior actions that nearly banned all uses of the product, several uses were authorized. The companies with products registered for use in interior paint agreed to voluntarily cancel their registrations in June 1990. The agency cited the voluntary cancellations, which also required special labeling of existing paint stock, as a success because the entire process took about six months to complete.

Hearing Process

Under certain circumstances, a registrant may request a hearing to review EPA's proposed risk management action under FIFRA Section 6. Hearings may be requested as part of a special review process, upon receipt of a notice of intent to cancel or suspend a pesticide or its use. The rules of practice for such a hearing are codified at 40 CFR 164.

Indemnification

In the 1988 amendments to FIFRA, Congress limited the circumstances under which companies receive indemnification (cash compensation) for canceled or suspended products. Historically, such payments were automatic for a number of pesticide companies. Under the revised law, farmers remain the only group entitled to automatic reimbursement from the federal government. The 1988 amendments also required pesticide sellers to reimburse buyers for the purchase price of suspended or canceled pesticide, unless the seller stated in writing that no refunds would be granted.

INTERNATIONAL TRADE RESTRICTIONS

FIFRA Section 17 restricts imports and exports, and is at least partly a result of Congress's concern that unregistered pesticides and their residues would appear in the U.S. food supply.

Imports

Pesticide import rule-making is the responsibility of the U.S. Customs Service, in consultation with EPA. Any item that violates FIFRA standards can be refused admission to the United States and, in some cases, destroyed. In general, Customs is responsible for stopping illegal imports, and EPA is in charge of evaluating products that Customs seizes.

EPA may request a sample of an imported pesticide to determine whether it is adulterated or misbranded. A product may be delivered to the importer pending an evaluation by EPA, in which case the importing firm may be required to post a bond for the value of the shipment.

Exports

FIFRA Section 17(a) imposed labeling requirements on U.S. companies that export unregistered pesticides. U.S. companies may export such pesticides without violating FIFRA if the foreign purchaser signs a statement acknowledging that the product is unregistered and cannot be sold in the United States. A copy of the acknowledgment statement must be sent to EPA, and the agency sends a copy to officials in the importing country. If a pesticide is manufactured for export only, the U.S. manufacturer must adhere to all standards to which the foreign purchaser is subject.

The label of the exported, unregistered product must be written both in the language of the importing country and in English. This bilingual labeling must enable an ordinary individual to read and understand the information. The label must bear the statement "not registered for use in the United States." A warning and caution statement, an ingredient statement, and the word "poison," if required, must appear in both languages. EPA issued a policy statement on exports in 1980, affecting 40 CFR 162–167, 169–172, and 180.

Pesticide exporters must keep records on the production directions and specifications of foreign purchasers for two years.

In 1990, EPA initiated a series of enforcement actions to call attention to the bilingual and other export labeling requirements. Six pesticide makers faced a total of about $1 million in proposed fines, reduced considerably after negotiations.

STATE AUTHORITY

Although FIFRA preempts most state regulation, state and federal cooperation is required in several areas. State authority is defined under Section 24. FIFRA allows states to:

- Administer applicator certification programs (see the discussion of these programs earlier in this chapter)
- Issue EUPs (see earlier discussion)
- Register pesticides to meet special local needs (see discussion below)
- Enforce federal pesticide laws and regulations (see the last section of this chapter)
- Regulate sale and use of a registered pesticide within a state, but only to the extent that such regulation does not

permit any sale or use prohibited under FIFRA (Section 24(a))

States cannot impose labeling or packaging requirements that vary from FIFRA standards (Section 24(b)). EPA must cooperate with states and other agencies to ensure uniformity of regulations (Section 22). The agency also may pay 50 percent of the costs of state training and certification programs (Section 23).

The Supreme Court clarified the extent to which FIFRA preempts local pesticide regulation in June 1991 (*Wisconsin Public Intervener v. Mortier*, 111 S. Ct. 2476, 33 ER Cases 1265 (1991)). The high court ruled on whether FIFRA preempted a local ordinance that required a permit from a town prior to application of pesticides to certain lands. The court concluded unanimously that FIFRA does not preempt such local rules.

Special Local Needs Registration

FIFRA also empowers states to register additional uses of federally registered pesticides to meet special local needs (Section 24(c)). Since the provision was enacted in 1972, thousands of special needs registrations have been granted. However, EPA retains the power to veto or suspend a state's authority. State-registered pesticides are subject to all provisions of FIFRA that apply to currently registered products (40 CFR 162).

EPA defines a special local need as an existing or imminent pest problem within a state for which the lead agency, based on satisfactory information, determines that an appropriate federally registered pesticide is not available. If a state approves a Section 24(c) registration, a copy of the application/notification must be forwarded to EPA for review within 10 days. The federal agency has 90 days to review it. The registration goes into effect immediately upon state approval, however. EPA periodically publishes a list of state-approved registrations in the *Federal Register*.

A state registration is only valid for distribution and use of a product within the state. States cannot authorize the use of a pesticide if EPA has canceled, suspended, or denied a registration for that use. States can issue new-use registrations only when the use is covered by a tolerance or exemption for pesticide residues as required by Section 408 of FFDCA. (See the complete discussion of pesticide tolerances earlier in this chapter.)

EPA has granted states some leeway for certain types of new products. States can issue registrations when the product is identical in composition to a federally registered product or contains the same active and inert ingredients in different proportions.

EPA also allows some older pesticides to be sold and used in intrastate commerce without federal registration. Such pesticides were registered before FIFRA preempted state registration authority. These sales and uses are authorized until EPA acts on the federal registration application (40 CFR 152, Subpart L).

ENFORCEMENT

Congress directed that states, with EPA's approval, assume primary enforcement responsibility for both civil and criminal violations under FIFRA Section 26. Most states now have EPA-approved FIFRA enforcement programs.

Because of the shared responsibility for FIFRA enforcement, lawsuits against violators may be filed in three forums. States with approved enforcement programs can file lawsuits in state courts. EPA may take administrative action through an EPA administrative law judge (ALJ). The federal courts handle all lawsuits involving criminal allegations, as well as challenges to decisions by ALJs.

FIFRA also empowers EPA, or an EPA-designated official, to conduct inspections of facilities (Section 9). If a violation is discovered, enforcement officials can file an administrative complaint seeking penalties (Section 14(a)). For minor violations, officials can file a notice of warning against the violator instead of seeking civil penalties. FIFRA Section 14(b) authorizes government officials to pursue criminal violations.

If a pesticide or device is found to be in violation of FIFRA, EPA may issue a "stop sale, use or removal order" to prohibit distribution of the product (Section 13). The government also has authority to seize (Section 13(b)) or obtain a court injunction (Section 16(c)) to stop distribution of an illegal product. The agency, in conjunction with the U.S. Customs Service, can refuse or detain an imported pesticide if it determines the shipment is not in compliance with FIFRA.

EPA's Enforcement Response Policy (ERP) for FIFRA, a 49-page document, outlines the agency's general enforcement procedures. One section of the ERP describes EPA's policy for negotiating settlements of civil violations. Another outlines civil penalty computations, including the penalty worksheet, the list of penalty adjustment criteria, and the criteria for determining the level of the penalty—the gravity-based penalty matrix for FIFRA.

EPA also has developed enforcement response policies for specific FIFRA programs, such as those governing good laboratory practices and establishment reporting requirements. These policy statements are modeled after the general ERP.

Violations

The following activities are prohibited by FIFRA Section 12:

- The sale of unregistered, adulterated, or misbranded pesticides (Section 3)
- The use of any registered pesticide in a manner inconsistent with its labeling (Section 25)
- The production of pesticides in an unregistered establishment (Section 7)

Before charging a company with a FIFRA violation, EPA or the state must first issue a warning notice.

In 1988, EPA issued rules governing pesticide advertising in an attempt to curb false or misleading claims (40 CFR 168). It is unlawful to advertise any pesticide with a use authorized only under:

- A FIFRA Section 5 EUP
- A FIFRA Section 18 emergency exemption, except when the advertisements are confined to the geographical areas covered by the exemption
- A FIFRA Section 24(c) special local need registration

Failure to properly file an annual pesticide production report also is subject to enforcement. Most cases alleging failure to file annual reports are initiated by EPA, not the state.

The government also may file complaints alleging violations such as failure to maintain records, distribution or sale of a product with improper coloring, removal or alteration of a pesticide label, disclosure of confidential business information, falsification of data, and violation of a suspension or cancellation order.

Penalties

Section 14 of FIFRA authorizes EPA to assess civil penalties of up to $5,000 for each violation by a registrant, commercial applicator, wholesaler, dealer, retailer, or other distributor. Section 14 provides for criminal fines as well.

A registrant, commercial applicator, wholesaler, dealer, retailer, or other distributor may be fined up to $5,000 for each offense. General-use applicators that violate any provision of FIFRA may be assessed $500 for the first violation and $1,000 for each subsequent offense. Private applicators may be assessed up to $1,000 for each offense.

No fines can be assessed if the person charged was not given a warning of noncompliance and an opportunity for a hearing. In determining the amount of the fine, EPA must consider the size of the business, the effect on the ability of the person to continue in business, and the gravity of the violation.

Congress authorized criminal penalties when a registrant, commercial applicator, wholesaler, dealer, retailer, or other distributor knowingly violates the law. Such violations are misdemeanors that are punishable by no more than one year in jail and/or $25,000 in fines. All other persons, including private applicators, who knowingly violate FIFRA are guilty of a misdemeanor and, on conviction, can be fined up to $1,000 and/or 30 days in jail.

Civil penalties may be waived if:

- A person provides a written assurance that the product met FIFRA's requirements at the time of sale
- Shippers or transporters allow EPA to copy all their records on pesticides
- A public official, while engaged in the performance of official duties, violates FIFRA
- A person is using a pesticide authorized by an EUP
- A person is shipping a pesticide for testing purposes to determine the product's toxicity and efficacy and does not expect to receive any benefit for the pesticide's use.

State Primacy

Most enforcement is carried out by state agencies, usually through a branch of the state agriculture department. Sections 26 and 27 of FIFRA outline the criteria states must meet to obtain primary enforcement responsibility and the conditions under which EPA can revoke state primacy, which EPA expanded in a 1983 rule (40 CFR 173).

States can obtain primacy to enforce pesticide use standards by adopting laws that are as stringent as federal requirements. States also can gain primacy through a cooperative agreement with the EPA administrator, in which case the agency is authorized to pay up to half the costs of the program. States with approved applicator certification programs also have primary enforcement authority for those programs.

EPA can step in and initiate an enforcement action if requested by a state or if a state does not begin enforcement proceedings within 30 days after notification of significant pesticide use violations.

Enforcement Examples

The following are examples of enforcement initiatives.

State Actions

In August 1990, the California Department of Food and Agriculture seized seven acres of table grapes on which the growth regulator 4-CPA was applied illegally. State law authorizes such seizures as well as fines of up to $25,000 per violation, although no fine was reported in this case.

In June 1990, ChemLawn Services Co. settled with New York over a suit brought in state court over its safety claims. The company agreed to stop making the claims and pay New York $100,000 in litigation costs. The settlement only covers New York, however, and ChemLawn did not admit to making false or misleading claims (*New York v. ChemLawn*, N.Y. Sup. Ct. NYCty, No. 40533/88).

The Washington Department of Ecology fined a homeowners organization $7,000 and a commercial herbicide applicator $18,000 in January 1991 for applying Aquacide to a lake without a permit. The product contains 2,4-D, a chemical that has been banned from state waters. The action was brought under the state water quality law.

Federal Administrative Proceedings

Administrative actions are filed by EPA when a state requests assistance or chooses not to litigate a violation. EPA also files enforcement actions for states that do not have primary enforcement authority. The federal government has exclusive enforcement responsibility for FIFRA Section 7(c) annual pesticide establishment reports.

Companies may negotiate a settlement with EPA directly or ask for a hearing before an ALJ. Negotiated agreements, also known as consent agreements, are made enforceable by accompanying consent orders approved by EPA's Environmental Appeals Board.

Administrative decisions are issued by ALJs. They may be appealed to the Environmental Appeals Board, a three-judge panel that replaced the agency's chief judicial officer in 1992. If EPA loses that appeal, the litigation ends. However, a company that loses such an appeal may ask a U.S. district court to review the decision. The following are examples of administrative decisions.

- *Export Provision:* Officials from the Monsanto Co. and EPA negotiated a settlement of alleged export violations that resulted in one of the largest fines ever collected under FIFRA. Monsanto agreed to pay $267,920 to settle charges that the company exported 175 shipments of unregistered products and failed to obtain a statement from the purchaser acknowledging that such products were not registered for use in the United States as required by Section 17(a) of FIFRA (*In re* Monsanto Co., FIFRA 90-H-8, April 2, 1991).
- *Efficacy Claims:* A $10,000 fine was levied against Sporicidin International for making unauthorized advertising claims about two of its disinfectants, and the decision was upheld on appeal by EPA's chief judicial officer. The company allegedly claimed that its products were effective against the AIDS virus. EPA charged the company with making false efficacy claims, distributing a product that was substantially different than the product registered with EPA, and shipping products that wre misbranded (*In re* Sporicidin International, FIFRA 88-H-02, June 4, 1991).
- *Illegal Sale of Restricted-Use Pesticides:* In 1985 EPA charged that Helena Chemical Co. sold restricted-use pesticides to uncertified applicators on 27 occasions and falsified pes-

ticide sales records. EPA proposed a fine of $135,800. EPA appealed an ALJ's decision against the agency to EPA's chief judicial officer, who in 1989 overturned the first decision and imposed a fine of $117,400 (*In re* Helena Chemical Co., FIFRA Appeal 87-3, Nov. 20, 1989).

- *Failure to File Annual Report:* EPA charged Cascade Chemical Inc. with failure to file an annual report of pesticide production for the year 1985 by February 1, 1986, as required by FIFRA Section 7(c). EPA proposed a fine of $3,200, but the fine was reduced by an ALJ to $800. EPA attorneys endorsed the penalty reduction because the company's annual gross revenues indicated the company was small in size (*In re* Cascade Chemical Inc., FIFRA 1086-03-04-012, Sept. 26, 1986).

- *Misuse of a Pesticide:* Bartlett and Co. Grain, a grain storage company, appealed an ALJ's decision that the company used a pesticide in a manner inconsistent with its label and failed to keep the required use record for a restricted-use pesticide. The ALJ assessed a civil penalty of $9,200. During an inspection, Bartlett's manager said he had recently used phostoxin to control insects in grain. The labels of the canisters indicated that the authorization period for the pesticide's use had expired and that gas masks, which were required by the labeling instructions, were not worn. The agency's chief judicial officer upheld the decision and penalty (*In re* Bartlett and Co. Grain, FIFRA Appeal 86-5, Nov. 25, 1988).

Federal Courts

The Department of Justice litigates lawsuits alleging civil or criminal violations of FIFRA in federal courts, even though EPA investigates and develops the case.

In 1988, Katzson Bros. Inc., a wholesale supplier of janitorial laundry and dry-cleaning products, sued EPA in a federal appeals court, claiming that a $4,200 fine levied by an ALJ was excessive. The court agreed that the fine was "extreme" and reduced it to $800. EPA had charged the company with failure to file a FIFRA Section 7(c) annual pesticide establishment report (*Katzson Bros. Inc. v. EPA*, 839 F. 2d 1396, 27 ER Cases 1425 (10th Cir. 1988)).

A laboratory owner who pleaded guilty to submitting false information to EPA in connection with a DCI was fined $1,000 and sentenced to 30 days in prison and one year of probation by a U.S. district court. Felix Konstandt, owner of Konstandt Laboratories Inc., also was convicted of one count of knowingly making false statements by providing 19 falsified results-of-analysis reports. For those violations of the general criminal laws, the laboratory owner was ordered to pay a $100,000 penalty under the Alternative Fines Act (*U.S. v. Konstandt Laboratories and Felix Konstandt,* Nos. 89-00563, 89-00676 (E.D.N.Y. Aug. 24, 1989)).

Chapter 2

RIGHT-TO-KNOW REQUIREMENTS

INTRODUCTION

After several tragic incidents around the world involving chemical spills, including the notorious release in Bhopal, India, in 1984 that killed more than 2,500 people, public pressure came to bear on the Environmental Protection Agency (EPA) and Congress to respond. In response, Congress passed the Emergency Planning and Community Right-to-Know Act (EPCRA) in 1986 (42 U.S.C. 11001, *et seq.*). The act was passed as Title III of the Superfund Amendments and Reauthorization Act, but was created as a free-standing law.

There are four basic sets of requirements that are imposed on federal and state authorities and industry:

- *Emergency Planning:* State and local governments must establish a framework for handling emergency releases and potential accidents. Industry has input into the planning, and each industrial facility is required to notify local officials if it handles certain chemicals.
- *Emergency Releases:* If there is a release of any one of over 700 chemicals defined as hazardous under the Superfund

law, the responsible facility must report the release if the amount exceeds a given reportable quantity.

- *Chemical Inventory:* Facilities are required to submit specific information on chemicals handled at the plant, culled from material safety data sheets (as already required by the Occupational Safety and Health Act) and from EPA reporting forms.
- *Routine Toxic Releases:* The most controversial part of the new law, this portion requires facilities to provide an annual list of releases and amounts of regulated chemicals. This information is then included in a computerized Toxic Release Inventory (TRI) database that is accessible by the public.

EPCRA has the largest number of facilities subject to its requirements—up to 4 million—of any federal environmental law, but as of July 1990 EPA officials were estimating that only half that number were complying with the reporting law. Even so, the agency has managed to produce informative statistics on chemical releases: EPA's review of 1988 data showed that 4.77 billion pounds of toxic chemicals were reported to have been released from the 20,000 facilities that reported that year, with Texas and Louisiana accounting for 24 percent of releases and 50 facilities accounting for 36 percent of all releases. An incomplete review of 1989 data showed total releases of 4.24 billion pounds, with Texas and Louisiana alone accounting for over 1 billion pounds of releases. With the lag in compliance, however, EPA officials threatened an aggressive enforcement program—counting on filing up to 400 actions in 1991 to enforce but one of the four key EPCRA provisions.

EMERGENCY PLANNING

Section 301 of EPCRA requires state and local governments to set up the basic infrastructure for dealing with emergency releases of toxic substances. While these governments exercise the main authority in developing emergency response and preparedness plans, industry and community groups are provided opportunities for input. Section 302 requires EPA to establish a list of substances and facilities covered, along with duties imposed for notice about handling such substances. Section 303 lays out criteria for local communities to use in developing comprehensive emergency response plans (42 U.S.C. 11001–11003).

State and Local Governments

The planning process begins when the governor of a state selects a State Emergency Response commission, which has the responsibility to establish local planning districts and appoint Local Emergency Planning committees. State duties also include establishing procedures for handling information that facilities must make public under EPCRA.

In turn, local committees must develop local emergency response plans for their jurisdictions. Members of the local committees will typically include local elected officials, fire or civil defense personnel, representatives from a local newspaper or television station, community group members, and spokespersons for regulated facilities.

Despite hopes for interaction among all these groups, a study submitted to EPA in August 1990 showed that local committees typically had no contact with local environmental groups, received no public requests for right-to-know information, and devoted minimal time to public education about hazardous materials. The study indicated that citizens are generally unaware of the existence of local committees, instead obtaining information from local government officials.

Substances and Facilities Covered

Reporting requirements apply to facilities that have certain types of chemicals present in quantities above thresholds set by EPA. Facilities are defined to include all structures and equipment "located on a single site or on contiguous or adjacent sites and which are owned or operated by the same person" (40 CFR 355.20).

These facilities must notify the state authority if they have what EPA has deemed to be "extremely hazardous substances" in levels at or above the threshold planning quantity set by the agency. The list of substances covers some 300 chemicals with threshold planning levels of from 1 pound to 10,000 pounds (40 CFR 355, Appendix A). The amount at stake for each facility includes the "total amount of any extremely hazarous substance present at any one time at a facility in concentrations greater than one percent by weight, regardless of location, number of containers, or method of storage."

Notification

The state commission must have been notified by May 17, 1987, or within 60 days after the facility becomes subject to these EPCRA requirements. In addition, the facility operator must designate a representative who will take part in the local committee's emergency planning activity (40 CFR 355.30).

Emergency Response Plans

The local committees had until October 17, 1987, to develop response plans for potential releases from these facilities; however, EPA later estimated that only about 50 percent of the 3,800 committees had met the deadline. The agency has recognized two key factors hindering full compliance: lack of knowledge and funding. In response, EPA established a peer exchange program for exchange of advice and information among committees, along with providing grant programs.

Local plans are required to identify:

- Facilities and potential routes of a chemical release

- Methods and procedures for local and facility personnel to follow if there is a release
- Procedures for providing adequate notice to the proper officials and the public about a release
- Ways to determine that a release has occurred, along with the extent of the affected area
- Emergency equipment available to respond
- Evacuation plans
- Training programs for local emergency personnel

EMERGENCY RELEASES

Section 304 of EPCRA ties together reporting requirements under both this law and the Comprehensive Environmental Response, Compensation, and Liability Act (CERCLA). If there is a release of any chemical listed under Section 302 or listed as a hazardous substance under Section 103(a) of CERCLA, facilities must report the release to local and state officials. For substances on the CERCLA list, which is almost identical to the one under EPCRA, the operator must also report the release to the National Response Center in Washington, D.C. Under both laws, reporting requirements are triggered if the release exceeds the reportable quantity established by EPA.

Facilities are defined by EPA under Section 304 in the same manner as under Section 302: "all buildings, equipment, structure, and other stationary items that are located on a single site or on contiguous or adjacent sites and which are owned or operated by the same person" (40 CFR 355.20). However, the reportable quantity limits are much lower. In short, reporting is required where a facility produces, uses, or stores a hazardous chemical and there is a release of a reportable quantity of a substance.

Exceptions apply (40 CFR 355.40(a)):

- When the release only exposes persons within the boundary of the facility
- For releases permitted under CERCLA Section 101(10), such as releases allowed under a Clean Water Act or Resource Conservation and Recovery Act (RCRA) permit (see related material in the discussion of CERCLA in Chapter 6)
- For releases that are continuous and stable in quantity and rate, unless there is a significant increase in release
- For releases of pesticides
- When the release does not meet the CERCLA definition of "release" under Section 101(22), such as emissions from motor vehicle exhausts
- For releases of certain types of radionuclides.

Reporting is triggered if the chemical release exceeds the reportable quantity established by EPA. Two sets of chemicals are listed by EPA: one for CERCLA hazardous substances (40 CFR 302.4) and another for extremely hazardous substances under EPCRA (40 CFR 355, Appendix A).

Typical examples of facilities and substances covered in this section include tanker trucks or railroad cars carrying hazardous substances, dairies that have to deal with discharges of ammonia, and power companies storing transformers containing polychlorinated biphenyls (PCBs). On a larger scale, one nationally recognized company suffered an explosion and fire at an Ohio facility that produced linings for cans and coatings for paper cups. Although the firm had complied with chemical inventory reporting requirements and had reported the incident within five hours, state environmental officials complained that the company missed a state-imposed 30-minute time limit for reporting the blast.

If a release of a reportable quantity of a substance occurs, the person in charge of the facility has the responsibility of notifying the appropriate authorities. Under CERCLA, immediate notice must be provided to the National Response Center ((800) 424-8802; in Washington, D.C. (202) 426-2675). In addition, the facility owner/operator must follow notice procedures specified by EPCRA.

Under EPCRA, the operator must immediately notify the community emergency coordinator and the State Emergency Response Commission (40 CFR 355.40(b)). If there is no local coordinator, the operator must contact other local emergency personnel (e.g., a fire department). As an alternative, the information may also be given to the emergency telephone operator in the area. The following information should be provided:

- Name of the chemical involved
- Whether it is an extremely hazardous substance
- An estimate as to the quantity released
- Time and duration of the release
- Media into which the release occurred (air, water, land)
- Health risks associated with the substance
- Proper precautions, such as the need to evacuate people
- Name and telephone number of the person at the facility to contact for further information.

As soon after the incident as possible, the operator must provide follow-up information, in written form, to state and local officials. Further information will be needed on potential health effects and suggested medical attention for exposed individuals, along with a description of actions taken in response to the release.

A brief example of the process is as follows: An Illinois ice cream manufacturer has an accidental release of ammonia from a refrigeration unit. The plant manager must quickly verify the type and amount of the release: since ammonia is a hazardous substance (40 CFR 302.4), a leak of greater than 100 pounds (45.4 kg) would trigger reporting requirements under EPCRA and CERCLA.

The plant operator must then telephone the National Response Center, the Illinois Emergency Response Commission, and the local committee that oversees facilities in the area. The information provided by the operator would include the elements mentioned above, except that ammonia is not an extremely hazardous substance. After initial notice, the plant operator must provide a written notice with further information about responding to the ammonia release.

ANNUAL CHEMICAL INVENTORIES

Tied to the community right-to-know concept of the law, Sections 311 and 312 of EPCRA require facilities to report information about chemicals handled at the site (33 USC 11021, 11022). Under Section 311, facilities that are required to prepare material safety data sheets (MSDS) under the Occupational Safety and Health Act (OSHA) must also provide such information to state and local authorities. Section 312 expands on the regulation by requiring more specific information on chemical inventories, based on EPA's reporting requirements (40 CFR 370).

The Hazard Communication Standard under the OSHA law identifies hazardous chemicals as those that pose a physical or health hazard (29 CFR 1910.1200). Although there is not a specific list of chemicals provided under the standard, the types of substances covered include those that pose cancer-causing, toxic, or corrosive hazards (e.g., solvents or chemicals such as benzene used throughout industry, and industry-specific chemicals such as methylene diphenyl di-isocyanate used in plastics manufacturing). Exemptions apply to:

- Hazardous wastes identified under the Resource Conservation and Recovery Act (RCRA)
- Tobacco products
- Wood products

- Certain articles with a specific shape or design in manufacture or end use (such as plastics) or that do not release hazardous chemicals in normal use
- Foods, drugs, cosmetics, or alcoholic beverages in retail establishments or that are for personal use of employees (similarly, items regulated by the Food and Drug Administration are excluded)
- Substances used for household or medical facility purposes
- Substances used for routine agricultural purposes

MSDS Reporting

An owner or operator of a facility must have submitted a report to the state commission, local committee, and fire department by October 17, 1987, if 10,000 pounds or more of a hazardous chemical is present at the facility. If extremely hazardous substances are used at a facility, the reporting threshold is 500 pounds, 55 gallons, or the threshold planning quantity for the substance, which is frequently less than the 500-pound/55-gallon amount for other extremely hazardous substances (see the listing at 40 CFR 355, Appendix A). If a facility has only begun using a substance, it has three months to meet reporting requirements.

For hazardous chemicals present in quantities of 10,000 pounds or less, an MSDS was required by October 17, 1989. The deadline for reporting new usage is 27 months.

Instead of providing an MSDS, the owner or operator can provide a list of hazardous chemicals for which a report would be required. Chemicals would be grouped by hazard category: (1) immediate (acute) health hazard; (2) delayed (chronic) health hazard; (3) fire hazard; (4) sudden release of pressure, as with compressed gas; and (5) reactive hazard. The listing also must identify hazardous components of each such chemical, as would normally be required on an MSDS.

If the owner or operator discovers new information about a hazardous chemical, updated MSDS information must be provided to authorities within three months.

Inventory Reporting

Facility owners and operators are also required by EPA to submit reports on chemical inventories (40 CFR 370.25). Quantities subject to reporting are the same as for MSDS reports, with different deadlines:

- For facilities handling at least 10,000 pounds of hazardous chemicals or 500 pounds of extremely hazardous chemicals, the initial deadline was March 1, 1988, although facilities that have only begun to be subject to the reporting requirements have until March 1 of the following year.
- For levels of hazardous chemicals between 0 and 10,000 pounds in the preceding calendar year, reports were due by March 1990, with a deadline of March 1 of the second year after a facility has newly become subject to the requirements. EPA has established Tier I and Tier II levels of reports, to be provided to the state commission, local committee, and fire department. Tier I reports are mandatory for all facilities, and Tier II reports may be requested by state or local authorities.

Tier I reports are less specific, requiring information on estimated maximum amounts of chemicals on-site (40 CFR 370.40). The information is not chemical-specific, instead requiring identification of physical and health hazards posed, such as fire, sudden release of pressure, reactivity, immediate (acute) toxicity, and delayed (chronic) toxicity. In addition, the owner or operator must describe the general location of the chemicals, along with the average daily amount and number of days on-site for that hazard group.

Tier II reports require more specific information on chemical types and CAS number for the substance (40 CFR 370.41).

Public Access to Reports

Anyone can obtain copies of MSDS reports by submitting a written request to the local committee. Although EPA has not provided for public access to Tier I reports, anyone can request Tier II information about a specific facility by submitting a written request to the state commission or local committee. For chemicals at a facility in amounts of less than 10,000 pounds, the request must also include a statement of need for that information. Lastly, a facility owner or operator can request that the location of any Tier II chemical be withheld.

TOXIC CHEMICAL RELEASE REPORTING

Perhaps the most controversial provision of EPCRA is the section requiring manufacturers to report releases of toxic chemicals, with reports made available in a database for relatively easy access by the public (42 USC 11013 or Section 313 of EPCRA). EPA has said that one of the purposes of the reporting requirements is to "inform the general public and the communities surrounding covered facilities about releases of toxic chemicals" (40 CFR Part 372). The TRI data are stored on the National Library of Medicine computer, and on-line retrieval costs run from $18 to $25 per hour. EPA has reported that industry compliance with TRI requirements is running at 67 percent.

TRI information has been used by individuals and environmental groups to track compliance with EPCRA and other environmental laws, verify assertions by industry about reductions in releases of toxic chemicals, and badger EPA about its follow-up on tardy or incomplete reports. State environmental authorities have also kept track of TRI data, using the information to design pollution prevention programs and to determine the need for further monitoring of facilities.

Facilities Covered

The facilities subject to reporting under Section 313 must have 10 or more employees and fall under primary

Standard Industrial Classification (SIC) codes 20–39, which cover industries such as food, textile, chemicals, petroleum, machinery, electrical equipment, and metals. In addition, the facility must have manufactured, imported, processed, or otherwise used a toxic chemical.

Chemicals, Thresholds

EPA has listed toxic chemicals to be reported, including more than 300 chemicals and 20 chemical categories (40 CFR 372.65). The reporting threshold is 25,000 pounds for chemicals manufactured or processed, and 10,000 pounds for chemicals that are put to other use at a facility. For companies manufacturing or using chemicals that are identified by category, the facility must consider combined production of chemicals that fit within the category to determine whether the threshold reporting amount is exceeded.

The chemicals covered in the reporting requirements frequently trigger attention from sources other than environmental agencies (e.g., regulation of chlorofluorocarbons under EPCRA has caused the Internal Revenue Service to review such data to ensure that facilities are paying proper excise taxes for ozone-depleting chemicals, as required by international law).

Chemical mixtures involve a more complicated process for determining threshold reporting amounts (40 CFR 372.30). In general, for mixtures or trade-name products the facility owner/operator must determine the amount of a toxic chemical in the mixture and add that to the amounts of the same chemical that are also manufactured, processed, or used at the facility.

However, if the owner/operator lacks specific information about chemical concentrations or chemical identity, information about the mixture or trade-name product may not have to be factored into the threshold quantity calculation. To make matters easier, suppliers of mixtures or trade-name products are required to notify customers about the presence and composition of Section 313 chemicals in their products (40 CFR 372.45).

Exemptions

As with other EPCRA provisions, certain exemptions have been provided by EPA:

- Toxic chemicals present in only a minimal amount. If a toxic chemical is in a concentration of less than 1 percent of a chemical mixture, or less than 0.01 percent for carcinogenic chemicals, then the amount will not count toward the threshold determination.
- Toxic chemicals in articles. Articles in this instance are defined in the same manner as above for annual inventories: items formed to specific shape or design during manufacture, or with an end use dependent on that shape or design, and that do not release a toxic chemical under normal processing or use.
- Toxic chemicals used in the structure of a facility, for routine janitorial or grounds use, by employees for personal use, for vehicle maintenance, or in cooling waters, compressed air, or air for combustion at the facility.
- Laboratory use of toxic chemicals.
- Owners of leased properties whose sole interest is the real estate.

Reporting

Reports on toxic releases are due July 1 each year, and are to be submitted to EPA and the state emergency response commission. Reports must cover releases that occurred in the preceding calendar year.

EPA has issued a four-part form to be completed for each chemical above the threshold at a facility. Form R must be filled out with the information below (40 CFR 372.85(a)). Note, however, that EPA has proposed to remove Form R and its instructions from the Code of Federal Regulations based on space limitations and changes in the form. Facility owners/operators should contact their regional EPA office for updated information (or the EPCRA hotline: (800) 535-0202).

- *Part I, Facility Identification Information:* (1) certification of accuracy by a senior management official; (2) identification of the facility, including SIC codes, Dun & Bradstreet number, and EPA identification number; (3) the number of any pollutant discharge permit the facility may have obtained under the Clean Water Act, along with a description of the water bodies that received discharges; (4) identification of any underground injection wells; and (5) parent companies of the facility.

- *Part II, Off-Site Locations to Which Toxic Chemicals Are Transferred in Wastes:* (1) publicly owned treatment works or sewage treatment facilities where waste is sent; (2) identification of any other off-site facilities, including facilities regulated under the RCRA (in addition, the owner/operator must indicate whether these off-site facilities are under the control of the reporting facility or a parent company).

- *Part III, Chemical-Specific Information:* (1) the chemical identity, based on the CAS number or a generic name if the identity is a trade secret (the owner/operator will have to prove the secrecy claim); (2) identification of components of mixtures; (3) description of categories of usage for the chemical; (4) maximum amount of the chemical on-site at any time during the year; (5) releases to the environment, identification of whether releases were to land, air, or water, and the amounts; (6) transfers of the chemical in waste to off-site sewage treatment or other facilities; (7) waste treatment methods and their efficiency; and (8) information about waste minimization efforts.

- *Part IV, Supplemental Information:* used only to provide additional information about facility identification, releases of a chemical to land, off-site transfer of chemical wastes, and waste treatment.

Pollution Prevention

In late 1990, the Pollution Prevention Act was signed into law, with new reporting requirements to be fulfilled by industry in Form R. In essence, the waste minimization information that was formerly optional is mandatory beginning with the July 1, 1992, reporting deadline. The new information required will include data on source reduction and recycling, estimates of changes in hazardous waste production over the next two years, details of production changes, and descriptions of programs used at the facility to encourage pollution prevention. Even with the new reporting, the law does not actually require industry to reduce the amount of hazardous waste. EPA is preparing regulations on the new reporting mandate.

TRADE SECRETS

For many of the reporting requirements under EPCRA, facility owners/operators are allowed to claim that confidential business information is at stake (42 U.S.C. 11042, 11043, 11046). According to EPA, a business may have "the right to preserve the confidentiality of business information and to limit its use or disclosure by others in order that the business may obtain or retain business advantages it derives from its right in the information." The claim can apply to:

- Changes to a facility reporting about an emergency response plan
- Information submitted in response to requests by a local emergency planning committee
- Chemical identity information in an MSDS
- Chemical information submitted in a Tier II emergency and hazardous chemical inventory form
- Chemical identity information for purposes of the TRI

To make the claim, in general, the person requesting protection will first have to submit the information to EPA, including the specific chemical identity at stake (40 CFR 350.5). For the last three categories above, two versions of the information must be submitted to the agency: one "unsanitized" version of the report containing full information and one "sanitized" version with the confidential information omitted. The sanitized version must still be sent to state and local authorities, which can make it available to the public.

The person making the claim will also have to substantiate it by (40 CFR 350.7, EPA Form 9510):

- Describing measures taken to protect the confidentiality of the chemical identity
- Showing that the information has not been disclosed to anyone else, with the exception of the government or an employee
- Listing government agencies to which the information has already been disclosed, and whether that disclosure contained a claim of confidentiality
- Explaining why a particular use of a chemical is a trade secret
- Indicating what harm would result from disclosure of the chemical identity
- Describing the extent to which information about the chemical is available in products, articles, or testing of environmental releases

Once the claim of confidentiality is made to EPA, the agency will not disclose further information until a public request is submitted (40 CFR 350.15). After receiving the request, EPA has 30 days to determine whether the claim is supported. If the agency decides that there is no trade secret at stake, the person requesting it will be told to either amend the information supporting the claim or appeal the agency's decision to the EPA general counsel. If trade secrecy is still rejected, the next step is an appeal to a federal district court.

One major exception to trade secret protection applies: health professionals are entitled to receive chemical information if they present a written request and the information is necessary for the purposes of diagnosis and treatment of persons exposed to the particular chemical (42 U.S.C. 11043).

ENFORCEMENT

Various levels of enforcement authority may be used by the federal government and EPA, based on the EPCRA provision at stake and extent of violation (42 U.S.C. 11045):

- *Emergency Planning (Sections 302 and 303):* Failure to comply with an order under these provisions can lead to a civil penalty, enforced in federal district court, of up to $25,000 per day.
- *Hazardous Release Reporting (Section 304):* EPA can assess administrative penalties of up to $25,000 per violation, with penalties of up to $75,000 for a second or subsequent violation. The penalty structure is identical if the United States goes to court to collect penalties. For knowing or willful violations, criminal penalties can range up to $25,000 and a two-year prison term, or $50,000 and a five-year sentence for second violations.
- *MSDS Reporting (Section 311):* Administrative or judicial penalties can range up to $10,000 per violation.
- *Tier I, Tier II Reporting, TRI Reporting (Sections 312 and 313):* Administrative and judicial penalties can be up to $25,000 for each violation.
- *Trade Secrets (Section 322):* A frivolous claim of a trade secret can result in an administrative or judicial penalty of up to $25,000 for each claim. On the other hand, a knowing or willful disclosure of such a secret can lead to criminal fines of up to $20,000 and a year in jail.

Another avenue for enforcement—by citizens and environmental groups—has been authorized by EPCRA (42

U.S.C. 11046). Citizens must provide 60 days' notice prior to the suit to the alleged violator, EPA, and state authorities and can be precluded from going forward with a suit if the agency has begun an administrative or civil enforcement action against the facility. Civil penalties may be obtained by the citizen, along with attorney's and expert witness' fees.

In terms of actual use of enforcement authority, EPA thus far has emphasized Section 313 TRI violations. As mentioned previously, the agency planned 400 enforcement cases in 1991 under this section, twice the number filed in 1990. Most cases are settled, with penalties obtained in 97 percent of the 1990 settlements; fines against individual violators ranged up to several hundred thousand dollars. One new approach by EPA is to tie penalty assessment to pollution prevention measures taken by a violator. For instance, EPA reached a settlement with one Ohio company where the company would pay a $35,250 penalty but spend an additional $100,000 to install pollution control equipment. In another case, the agency negotiated an agreement whereby a civil penalty would be reduced by $1 for every $3 the Idaho manufacturer spent on environmentally beneficial projects.

Chapter 3

MANUFACTURING CONTROLS

I. CLEAN AIR ACT

INTRODUCTION

When Congress approved the Clean Air Act Amendments of 1990, it tackled three major problems that the original 1970 statute proved unable to solve: acid rain, urban ozone, and toxic air pollutants. It left in place, however, the three basic elements of the air pollution regulatory system that had proven effective: national ambient air quality standards (NAAQS) for the six major pollutants, new source performance standards for factories and power plants, and the mobile source program.

Overlaying all this is a new national air pollution permits program modeled after the successful permit program under the Clean Water Act. The National Pollutant Discharge Elimination System (NPDES) requires all dischargers into U.S. waters to obtain permits. This seamless regulatory scheme has resulted in significant improvements in water quality.

Congress also developed a new strategy to improve air quality: the law encourages automobile manufacturers to develop vehicles that run on fuels other than gasoline. The bill pushes the petrochemical industry to develop and market alternative fuels as well. Accompanying those incentives are tighter emission standards for conventional vehicles.

In addition, Congress expanded EPA enforcement authority, making it easier to issue administrative penalties and seek criminal prosecutions.

NATIONAL AMBIENT AIR QUALITY STANDARDS (NAAQS)

For regulatory purposes, EPA divides air contaminants into two broad categories: criteria pollutants and hazardous air pollutants. In both cases, EPA follows a single pattern. It develops numeric standards that apply nationwide and delegates to states the authority to enforce the standards and to administer the programs on a day-to-day basis.

Section 108 of the Air Act defines criteria pollutants as those "which may reasonably be anticipated to endanger public health or welfare." They might more accurately be described as "everyday, garden-variety air pollutants" whose names most people recognize and that most people inhale, in small quantities, every day. Research indicates that they do not cause cancer in humans, although they can cause other health problems.

The six commonly occurring air pollutants that fall into this category are carbon monoxide, lead, nitrogen dioxide, ozone, particulates, and sulfur dioxide. EPA sets NAAQS for each pollutant. These pollutants are discussed in greater detail below.

Theoretically, all sources of criteria pollutants that emit levels above a certain threshold are covered. Federally developed new source performance standards (NSPS) cover all new stationary facilities that emit any of the six criteria pollutants, regardless of where they are located. They also cover facilities that undergo major modifications.

Existing sources of criteria pollutants are regulated by the states in two ways. For areas that do not meet national ambient air quality standards, states must develop state implementation plans (SIPs) that will bring them into compliance. For areas that already comply, states must adopt prevention of significant deterioration (PSD) programs to ensure that new sources do not emit amounts that push an area into noncompliance.

State and federal regulators divide sources of criteria pollutants into three broad categories: stationary sources (such as factories and utilities), mobile sources (cars, trucks, buses, off-road vehicles, trains, airplaines), and area sources (small, numerous items, mostly consumer products such as paints and glues). While the first two categories have been regulated for years, area sources are only recently being addressed, primarily in California.

Carbon monoxide, lead, particulates, and sulfur dioxide typically are emitted directly from smokestacks and tailpipes. Controls generally are imposed on the largest sources—and automobiles—when air quality in an area exceeds NAAQs.

Ozone, on the other hand, is formed when hydrocarbons and nitrogen oxides are exposed to sunlight; most ozone control programs focus on limiting hydrocarbon emissions, although control of nitrogen oxides is being considered in certain areas.

Nitrogen dioxide can be emitted directly or can form through oxidation of other nitrogen oxides.

Under the Clean Air Act, EPA is authorized to set two types of standards: primary NAAQS designed to protect human health and secondary NAAQS designed to protect the environment. In practice, however, EPA has used a single standard for each pollutant as both the primary and secondary NAAQS.

More detailed information on criteria pollutants and compliance with ambient air quality standards can be found in EPA's annual *National Air Quality and Emissions Trends Report*. Detailed EPA rules on NAAQS can be found in Section 40, Part 50, of the Code of Federal Regulations.

Carbon Monoxide

This colorless, odorless gas forms during incomplete combustion of hydrocarbon-based fuels. More than two thirds of the carbon monoxide in the atmosphere comes from automobiles and other vehicles.

The carbon monoxide molecule is inherently unstable, carrying a positive electrical charge. It tends to combine with oxygen to form carbon dioxide, which is stable. Carbon monoxide attacks the respiratory system, robbing it of oxygen, and people with cardiovascular problems are most at risk. At high levels, it is extremely toxic.

The NAAQS for carbon monoxide includes two limits: a one-hour standard of 35 parts per million (ppm) and an eight-hour standard of 9 ppm. Nationally, carbon monoxide levels declined 32 percent from 1977 to 1986, according to EPA data. Agency officials predict that the decline will continue and that only a few cities will experience long-term carbon monoxide problems.

Lead

This pollutant is emitted by lead smelters, industries that use lead such as battery manufacturers, and engines that burn leaded gasoline. It also can be emitted by incinerators that burn waste containing lead.

In 1989 cars, trucks, and other vehicles emitted 2.2 million metric tons of lead, industrial processes contributed 2.3 million tons, and waste combustion accounted for another 2.3 million tons.

Lead's damaging effects on the human nervous system are well documented. It is considered particularly harmful to young children, and studies have suggested that early exposures can have long-term impacts.

The NAAQS for lead is 1.5 micrograms per cubic meter of air, averaged over a calendar quarter. Control of airborne lead represents one of EPA's greatest successes. The introduction of unleaded gasoline in 1975 and the significant reduction in lead levels in regular gasoline have helped cut ambient levels by 87 percent and cut emissions by 94 percent, based on a recent EPA assessment of 82 urban sites. Controls on smelters and battery plants also have contributed.

Nitrogen Dioxide

This pollutant and its relatives—nitric oxide, nitrous oxide, and others—are generated largely by motor vehicles and other types of fuel combustion; these sources accounted for 18.5 million of the 19.3 million metric tons emitted in 1986. Nitrogen dioxide is emitted directly and also forms during oxidation of nitric oxide. Its yellow-brown hue is partly responsible for the color of urban smog. Nitrogen dioxide is not by itself particularly harmful to humans, but it impairs visibility and is a precursor to ozone formation (see below). In addition, it is a secondary cause of acid rain. Regulators and air quality planners often use the term "NOx" to describe a mixture of nitrogen oxides. The "x" is used because it is not always known exactly what proportions of nitrogen dioxide (NO_2), nitric oxide (NO), nitrous oxide (N_2O), and a fourth compound, N_2O_3, are present at any given time. Their relative concentrations are determined by temperature, pressure, and humidity. For measurement purposes, regulators oxidize samples so that all of the NOx is transformed to NO_2, which is then measured.

The NAAQS for nitrogen dioxide is 0.053 ppm, measured as an annual average. Virtually all cities have met the NO_2 standard, and emissions have declined 8 percent from 1977 to 1986.

Ozone

This most persistent of pollutants is formed when hydrocarbons and nitrogen oxides interact in the presence of sunlight, and it is typically a problem in warm weather. An ozone molecule consists of three oxygen atoms and is inherently unstable; oxygen molecules normally contain just two atoms. Although it is commonly lumped together with other ingredients of "urban smog," it is invisible.

Ozone is a lung irritant and is of special concern to sensitive populations, such as the very young, the elderly, and individuals with asthma or other respiratory ailments. It also can affect individuals who are exercising heavily, and effects have been detected at relatively low levels. At this point, research has not identified any permanent effects, although there is speculation that ozone exposure could increase susceptibility to cancer.

The NAAQS for ozone is 0.12 ppm, measured as a one-hour average. Areas are allowed three days of "exceedances"—any one-hour average concentration that exceeds the standard—over a three-year period, and the fourth exceedance constitutes a violation. Although ozone levels are measured on an hourly basis, exceedances are tallied daily; thus, several one-hour exceedances in one day would represent, for compliance purposes, a single violation.

Ozone has proven the most difficult of the criteria pollutants to control, and more than 60 U.S. cities remain in violation of the standard. The primary control method has been to limit hydrocarbon emissions, but this has proven ineffective in many cities. Some areas are turning to NOx emission limits in an effort to reduce ozone levels.

Ozone is a problem in the summer for two reasons. First, the hotter weather leads to more evaporation of hydrocarbons, one of the precursors. Second, the longer and more intense sunlight provides more of the solar energy that causes the ozone-forming reaction.

Compliance with the ozone standard is based on a three-year time span. To compensate for unusual weather conditions and other factors, the law allows three exeedances in that period. As noted earlier, the fourth exceedance constitutes a violation.

The level of the fourth-highest exceedance is known as an area's "design value." The Clean Air Act divides areas into attainment categories, and cities with higher design values are required to do more to control ozone.

Particulates

These are any small particles that can be suspended in the atomsphere, but EPA is most concerned about fine particulates, which are thought to penetrate deepest into the lungs. EPA has abandoned the practice of measuring total suspended particulates (TSP) in favor of measuring only those smaller than 10 microns.

Fine particulates are emitted primarily by industrial processes, diesel-fueled vehicles, and other types of fuel combustion, including wood stoves.

The primary human health impact of fine particulates is lung cancer. Fine particualtes composed of various hydrocarbons are easily inhaled, whereas larger particulates generally are trapped and expelled by the respiratory system before they can cause severe damage.

The NAAQS for fine particulates is 50 micrograms per cubic meter of air measured as an annual mean, and 150 micrograms per cubic meter measured as a 24-hour mean. Compliance with the new standard is estimated by EPA based on old TSP measurements, and counties are classified by their probability of violating the new standards. Those with 95 percent or greater probability are classified as Group I and must develop new control plans. Those with a probability between 20 percent and 95 percent are in Group II and must perform monitoring. Those counties below 20 percent—Group III—are presumed to be in compliance. As with other criteria pollutants, these controls are managed through state implementation plans.

Long-term ambient trends are difficult to identify because of a change in measurement methods in 1981 and the shift to fine particulate control, but EPA estimates that emissions have declined 25 percent since 1977. Most gains have come in the industrial sector, however. New controls on diesel-fueled trucks and buses are expected to reduce emissions from transportation, which accounted for 1.4 million metric tons of the total 6.8 million tons of particulates emitted in 1986.

Sulfur Dioxide

This pollutant is primarily generated during combustion of coal and other fuels, accounting for 17.2 million of the 21.2 million metric tons emitted in 1986.

The health effects of sulfur dioxide are not considered especially severe, although moderate concentrations are thought to affect sensitive individuals, such as asthma sufferers.

There are three NAAQS for sulfur dioxide: an annual standard of 0.03 ppm, a 24-hour standard of 0.14 ppm, and a three-hour standard of 0.50 ppm.

Ambient levels of sulfur dioxide have declined 43 percent since 1977 according to one EPA sample, and 45 percent in another sample. The agency estimates that the number of violations of the NAAQS has declined 98 percent in that time.

New emission limits for coal-fired power plants may be required by Congress to control acid rain. Sulfur dioxide emissions are the principal cause of acid rain, and such emissions can be carried hundreds of miles by prevailing winds. Acid rain can increase the acidity of soil—killing trees—and water bodies—killing fish and other organisms.

NEW SOURCE PERFORMANCE STANDARDS

New facilities that emit pollutants are subject to NSPS requirements regardless of their location. These standards also apply to facilities tht undergo major modifications. NSPS requirements are somewhat unique in that they take effect as soon as they are proposed; otherwise, companies could evade the standards by constructing facilities in the interim period before final rules are in place.

NSPS requirements are designed to bring about attainment with ambient air quality standards and generally set limits on emissions of one or more of the six criteria pollutants. In the case of ozone, however, limits may also be imposed on emissions of volatile organic compounds (VOCs), which, along with nitrogen oxides, are precursors to ozone formation. A broad range of NSPS limits applies to industries such as petroleum refining, gasoline loading, synthetic organic chemical manufacturing, and various surface coating operations.

NSPS requirements do not specify what control equipment must be used, only the emission rate the source must meet. This means, for example, that a new coal-burning power plant could control sulfur dioxide emissions by using "scrubbers" to remove the pollutant, by using low-sulfur coal, by using natural gas, or by applying one of several emerging "clean coal" technologies.

Ironically, a power plant cannot meet the standard (1.2 ppm British thermal units of heat input) by using energy conservation, because that would not reduce the *rate* of emissions. It would, however, help control the ultimate problem: total loading of sulfur dioxide in the atmosphere, which determines how much acid rain will fall in the United States and Canada.

EPA has been criticized for not issuing a sufficient number of regulations or not doing so in a timely manner. This criticism does not apply to the NSPS program, however; since 1971, the agency has issued standards for 63 separate source categories. The following is a list of the sources regulated to date (some subcategories are not listed):

- Ammonium sulphate manufacturing
- Asphalt concrete plants
- Asphalt roofing manufacturing
- Automobile and truck surface coating
- Beverage can coating
- Brass and bronze ingot production plants
- Bulk gasoline terminals
- Coal preparation plants
- Dry cleaners

- Equipment leaks in petroleum refineries
- Equipment leaks in synthetic organic chemical manufacturing
- Ferroalloy production plants
- Fiberglass insulation processing
- Fossil-fuel fired steam generators
- Gas turbines
- Glass manufacturing plants
- Grain elevators
- Incinerators
- Industrial-commercial-institutional steam generating units
- Kraft pulp mills
- Large appliance surface coating
- Lead-acid battery plants
- Lead smelters
- Lime manufacturing plants
- Magnetic tape coating
- Metal coil surface coating
- Metal furniture surface coating
- Metallic mineral processing
- Nitric acid plants
- Nonmetal mineral processing
- On-shore natural gas processing
- Oxygen process steelmaking facilities
- Petroleum refineries
- Petroleum refinery wastewater systems
- Petroleum storage vessels
- Phosphate fertilizer plants

- Phosphate rock plants
- Portland cement plants
- Primary aluminum reduction plants
- Primary copper smelters
- Primary lead smelters
- Primary zinc smelters
- Residential wood heaters
- Rotogravure printing
- Rubber tire manufacturing
- Sewage treatment plants
- Steel plants
- Sulfuric acid plants
- Surface coating of plastic business machine parts
- Synthetic fiber production
- Vinyl and urethane coating and printing

EPA rules on NSPS can be found in Section 40, Part 60, of the Code of Federal Regulations.

MEETING THE STANDARDS

State Implementation Plans

An SIP is required for each air quality control region (or portion thereof) that exceeds an NAAQS.

Section 107 of the Air Act authorizes the EPA administrator to "designate as an air quality control region any interstate area or any major intrastate area which he deems necessary or appropriate for the attainment and maintenance of ambient air quality standards."

Section 110 of the Air Act states that the EPA administrator "shall, within four months after the date required for submission of a plan . . . approve or disapprove such plan or any portion thereof." The principal criterion for approval is that the plan provides for attainment of the standard "as expeditiously as practicable but . . . in no case later than three years from the date of approval. . . ." Revisions to previously approved plans are subject to similar requirements. States are required to provide reasonable notice and public hearings for SIPs or SIP revisions.

States may rely on a variety of measures to bring an area into compliance. They must, however, implement all measures that EPA has defined as "reasonably available control technology" (RACT) for individual pollutants. For example, if an area fails to meet carbon monoxide standards, it generally must impose an automobile inspection-and-maintenance program if it does not already have one.

EPA's determination that RACT exists for a given source is generally accompanied by "control technique guidance" (CTG) that describes how the reductions may be achieved. Some state and local officials, along with environmentalists, have complained that EPA has not made enough RACT and CTG determinations, forcing them to make the politically unpopular decisions themselves.

One of the ways that industrial facilities have tried to reduce the burden of Air Act requirements is to convince state authorities to revise their state implementation plans. If the operator of the plant can convince the state that its emissions control practices are more effective than otherwise required by the SIP, the state regulator may submit a revision to EPA for approval

At one time, this strategy was considered a way to stave off enforcement efforts by the federal government—if a facility operator thought that the plant was a likely candidate for enforcement action, a request might be made for an SIP revision. The logic was that EPA usually took an extremely long period of time to rule on such proposals, thus delaying the potential administrative or civil enforcement action.

However, in June 1990 the U.S. Supreme Court laid this tactic to rest. In a ruling on an SIP revision proposal applying to a General Motors plant in Massachusetts, the court gave EPA wide authority to act. The court first stated that there was no four-month time limit for EPA to act on an SIP revision, rejecting the claim that the four-month deadline for reviewing an original SIP did not apply to plan revisions. The more telling blow to industry was the holding that EPA may take enforcement action even if it delays unreasonably in reviewing an SIP revision. At best, according to the court, the offending company could ask for a reduction in sanctions if it could show how the delay harmed it (*General Motors Corp. v. U.S.*, 110 S.Ct. 2528, 31 ER Cases 1441, (1990)).

EPA rules on state implementation plans can be found in Section 40, Parts 51, 52, 62, and 81, of the Code of Federal Regulations.

The New Compliance Regime

Before Congress passed the Clean Air Act Amendments of 1990, regulators were basically forced to lump all noncomplying areas into one category: nonattainment. The new law divides violators of ozone and carbon monoxide standards into separate categories according to the severity of pollution.

For ozone, the law divided areas into five categories:

Area Class	**Ozone Level (ppm)**	**Attainment date** (from date of enactment)
Marginal	0.121 to 0.138	3 years
Moderate	0.138 to 0.160	6 years
Serious	0.160 to 0.180	9 years
Severe	0.180 to 0.280	15 years
Extreme	0.280 and above	20 years

Los Angeles is currently the only city in the extreme category. The severe category is divided into two groups, so that areas with ozone levels above 0.19 ppm would have 17 years to comply.

Areas must adopt various control measures based on their classification. Marginal areas must conduct comprehensive inventories of emissions and adopt any controls that were designated as RACT before passage of the law. The primary RACT is an automobile inspection and maintenance program. Moderate areas must adopt the same controls as marginal areas; in addition, they must demonstrate a 15 percent reduction in emissions of volatile organic compounds within six years of enactment. They also must control refueling emissions at gasoline service stations.

Serious areas must fulfill the same requirements as moderate areas, as well as enhanced vehicle inspection programs and enhanced monitoring. They also must adopt clean-fuel programs and transportation control plans.

Severe areas must develop plans to reduce the number of vehicle miles traveled, in addition to all requirements of serious areas.

Extreme areas must adopt even more stringent traffic control and clean-fuel programs, as well as all of the above requirements. Failure to fulfill these requirements or inability to reduce pollution levels will result in an area being moved into the next most severe category.

Other provisions are more stringent as the severity of pollution increases. The definition of "major source" for the purposes of stationary source controls declines from 100 tons of VOCs annually for marginal and moderate areas to 50 tons per year for serious areas, 25 tons per year for severe areas, and 10 tons per year for extreme areas. This is likely to encourage industry, when building new plants, to control emissions as tightly as possible to prevent a plant from being categorized as a major source. It also will prevent some high-emitting industries from locating in areas classified as serious, severe, or extreme.

If a company still chooses to locate a major new source in a non-attainment area, it must meet offset requirements. These grow more strict with each category. To site a new source of VOC emissions in a marginal area, planners must reduce emissions elsewhere, through shutdowns or tighter controls, on a 1.1:1 ratio. The ratios are 1.15:1 in moderate areas, 1.2:1; in serious areas, in severe areas, 1.3:1; and 1.5:1 in extreme areas. Congress adopted the offset requirements so that industrial growth could occur without exacerbating air quality. In fact, offset requirements ensure a net *decrease* in total emissions whenever a major new source begins operating. They give companies an economic incentive to install modern pollution control equipment in unregulated plants and to close inefficient, high-emitting plants.

Prevention of Significant Deterioration

The PSD program is designed to prevent an industry from moving into an area that meets air quality standards and emitting high levels of pollutants that eventually could cause violations of NAAQS. Some business representatives consider it extremely cumbersome, even Byzantine.

For each pollutant, EPA develops "increments," or portions of air quality standards that one pollution source can consume. For example, EPA issued, on October 17, 1988, final rules on PSD increments for nitrogen oxides. The NAAQS for NOx is 100 micrograms per cubic meter, measured as an annual average. EPA set the PSD increment at one-fourth that, or 25 micrograms per cubic meter. This means that no single company can emit NOx such that ambient levels increase by more than 25 micrograms per cubic meter.

Another requirement of the program, and the one businesses probably like the least, is that companies wishing to build new plants in such areas must conduct an entire year of baseline monitoring before the plants can begin operation. This establishes the ambient level of pollutants before a plant begins operation. The baseline monitoring is vital for regulators, however, because it enables them to calculate with some accuracy whether a plant has indeed increased pollutant levels beyond its increment.

MOBILE SOURCES

NSPS and the PSD program apply only to stationary sources. Many of the air pollution problems cities face are caused by motor vehicle emissions. Cars and trucks emit all of the criteria pollutants but ozone, and the nitrogen oxides and hydrocarbons they emit are largely responsible for ozone formation.

EPA has a highly developed vehicle testing and certification program based on standards that new vehicles must meet.

Automobiles are not only required to meet standards when they are new. They must meet them throughout their useful lives. EPA does its own "in-use" testing of older cars. It also requires areas with ozone and carbon monoxide pollution problems to adopt vehicle inspection-and-maintenance (I&M) programs. These programs identify vehicles whose emission control equipment is failing prematurely, as well as those whose owners have removed the devices or otherwise tampered with them. I&M programs are not popular but have become a permanent part of many urban vehicle regulation programs.

In the Clean Air Act Amendments of 1990, Congress recognized the dilemma facing automakers that seek to ensure compliance throughout the useful life of a vehicle. Legislators wanted to require all cars to meet emission standards for 10 years or 100,000 miles, but that threatened massive recall efforts years after the cars left the factories.

The solution was a two-tiered compliance regime. Cars and light trucks now must meet one set of emission standards for the first five years or 50,000 miles, and somewhat less stringent standards for their sixth through tenth years or up to 100,000 miles. Heavy-duty trucks would need to meet the second-tier standards for 11 years or 120,000 miles.

EPA's testing laboratory in Ann Arbor, Michigan, measures new vehicle emissions under simulated driving conditions. Tailpipe emissions of the criteria pollutants and evaporative emissions of hydrocarbons are included.

The Ann Arbor laboratory also calculates automobiles' fuel economy for the Department of Transportation (DOT). The two functions are combined at one facility in the interest of cost and efficiency. DOT is responsible for setting corporate average fuel economy (CAFE) standards under authority of the Energy Policy and Conservation Act, not the Clean Air Act.

Exhaust Emissions

All gasoline-powered motor vehicles must be equipped with catalytic converters, which convert carbon monoxide to carbon dioxide and various nitrogen oxides to nitrogen. The Clean Air Act Amendments of 1990 require that these and other emission control devices operate more effectively and over a longer portion of a vehicle's useful life.

These standards also will be phased in. For cars and light trucks, 40 percent of a manufacturer's production must meet the standards in model year 1994, 80 percent in 1995, and 100 percent thereafter. This covers particulates, hydrocarbons, carbon monoxide, and nitrogen oxides. For light trucks, however, the particulate requirements are pushed back one year: 40 percent in 1995, 80 percent in 1996, and 100 percent thereafter.

There are three basic categories of light-duty vehicles: cars and trucks that weigh up to 3,750 pounds, light-duty trucks that weigh between 3,751 pounds and 5,750 pounds, and light-duty trucks that weigh more than 5,750 pounds. The last category gives rise to the incongruous term "light-heavy trucks." All emission limits for these vehicles are expressed in grams per mile. The term "NMHC" refers to nonmethane hydrocarbons. Although methane is a hydrocarbon, its emissions are not considered hydrocarbons for testing and certification purposes.

Pollutant	**Vehicle Weight (pounds)**	**Certification 5 yrs/50,000 mi**	**Certification 10 yrs/100,000 mi**
NMHC	0–3,750	0.25	0.31
	3,751–5,750	0.32	0.41
CO	0–3,750	3.4	4.2
	3,751–5,750	4.4	5.5
NOx	0–3,750	0.4	0.6
	3,751–5,750	0.7	0.97
Diesel NOx	0–3,750	1.0	1.25
	3,751–5,750	—	0.97
Particulate matter	0–5,750	0.08	0.10

Source: Environmental Protection Agency

Light-duty trucks weighing more than 5,750 pounds must meet separate standards. In addition, trucks weighing between 3,751 pounds and 5,750 pounds must meet emission limits for 11 years or 120,000 miles. Note that there is some overlap in the two charts.

Pollutant	Vehicle Weight (pounds)	Certification 5 yrs/50,000 mi	Certification 11 yrs/120,000 mi
NMHC	3,751–5,750	0.32	0.46
	5,751+	0.39	0.56
CO	3,751–5,750	4.4	6.4
	5,751+	5.0	7.3
NOx	3,751–5,750	0.7*	0.98
	5,751+	1.1*	1.53
PM	3,751–5,750	—	0.10
	5,751+	—	0.12

*Diesel-fueled light trucks are not required to meet these standards.
Source: Environmental Protection Agency

Heavy-duty trucks are treated separately, for two reasons. First, they all have diesel engines, whose emissions are fundamentally different from those of the gasoline engines used in virtually all light-duty vehicles. Second, only a handful of large companies produce these diesel engines, but many smaller manufacturers use them in various configurations of trucks, buses, construction and earth-moving equipment, farm vehicles, and the like. In the interest of efficiency, EPA tests the engines themselves, not the vehicles. Emissions are measured in grams per unit of power, expressed as a "brake-horsepower-hour," or BHP hour.

Truck engines manufactured since 1977 are limited to 40 grams per BHP hour of carbon monoxide and 16 grams per BHP hour of hydrocarbons plus oxides of nitrogen. These last two pollutants are aggregated because control of one generally leads to increased emissions of the other.

Heavy-duty diesel engines made since 1977 are held to the same standards as their gasoline-powered counterparts. Diesel engines also must meet particulate standards. Starting with the 1991 model year, urban buses will be subject to limits on particulate emissions of 0.1 gram per BHP hour and a nitrogen oxides standard of 5 grams per BHP hour. Trucks have a less stringent 1991 particulate standard of 0.25 gram per BHP hour, but that will be tightened to 0.1 gram per BHP hour in 1994.

Modern gasoline engines typically emit extremely low levels of particulates, thus compliance is not an issue.

Engine manufacturers have not yet completely solved the diesel particulate emission problem. One solution is the particulate trap (essentially a filter), but it has drawbacks. It can clog or fill up and must be purged periodically, probably by burning off the trapped particulates; however, this creates safety and durability problems.

Another potential solution is a change to the engine itself to eliminate particulate emissions. Manufacturers are working on this, but the trick is to do so without sacrificing performance or increasing emissions of other pollutants.

Finally, manufacturers are looking into alternative fuels. Diesel engines can run on alcohol fuels, for example, but such fuels are not yet widely available. Fuel switching seems unlikely to solve the particulate emission problem in the short term, although it may be an important factor in the future.

Evaporative Emissions

When a motor vehicle is driven long enough to warm up and then is turned off, the engine generally remains quite hot; cooling system water no longer circulates, and there is no air flow to provide cooling. Thus, the fuel remaining in the carburetor and fuel lines "cooks," and much of it evaporates. These evaporative emissions contribute to ozone formation and contain substances that may be hazardous, such as benzene and toluene. To limit such emissions, modern automobiles have a sealed fuel system connected to a canister filled with activated charcoal, which traps and reclaims the evaporative emissions.

Regulators now must control another evaporative emission problem, related to refueling. When a gas tank is partly empty, the airspace is saturated with gasoline vapors. When a new supply of gasoline is pumped in, it displaces those vapors, which are forced out into the atmosphere. In California and the District of Columbia, fuel pumps must be equipped so that they reclaim those vapors; other states are in the process of adopting similar requirements.

The controls on gasoline pumps are commonly referred to as "Stage II" requirements. "Stage I" refers to comparable controls on bulk gasoline delivery—controls that have been in effect and accepted by the industry for years.

The alternative to Stage II control on gasoline pumps is equipment installed on the vehicle itself, commonly known as "on-board" vapor recovery control. This involves enlarging the existing charcoal canister to accommodate the refueling emissions and fuel lines between the gas tank and the canister. Condensed fuel from the canister is returned to the gas tank.

A long and bitter debate between the automobile and petroleum industries over the merits of each control method ensnarled Congress and EPA. DOT even got involved when various groups raised questions over the safety of on-board equipment in crashes.

Congress settled the matter in the Clean Air Act Amendments of 1990 by ordering automakers to install on-board equipment on 40 percent of all new cars within four years, 80 percent of cars within five years, and 100 percent of all new cars after six years. They must eliminate at least 95 percent of the emissions.

Any stage II requirements that had been in force will no longer apply to areas designated as moderate for ozone once EPA issues standards for the on-board controls. EPA also is empowered to waive or revise stage II requirements for areas designated as serious, severe, or extreme for ozone. Marginal areas are not required to install stage II controls.

RECALLS

If EPA discovers that any vehicle category is not meeting standards, the agency is empowered to order a recall.

The manufacturer then must repair or upgrade the emission control equipment without charging the customer.

Despite vast improvements in emission control equipment, recalls remain fairly common. In January 1991, EPA ordered Ford Motor Co. to recall more than 60,000 pickup trucks and vans that it said exceeded federal standards for hydrocarbons and evaporative emissions. The recall was based on routine in-use testing. Ford was required to install a new evaporative emission control canister and modify vacuum lines that regulate air flow to the catalytic converter.

ACID RAIN CONTROL

The Problem

The problem of acid rain is partly environmental and partly political. It has been documented by EPA's National Acid Precipitation Assessment Program that sulfur dioxide emissions lead to acidified rainfall, which in turn creates acid conditions in soil and water bodies.

Much of the research on acid rain involves damage in high-altitude forests. Acidified soil can attack a tree's root system, and eventually the leaves turn yellow, the outer branches become brittle, and the tree dies. By the time these exterior symptoms are visible, the root system has been largely destroyed.

Acidification of water bodies also has drawn much attention. Such acidification can impose direct stress on fish and can diminish populations of invertebrates and other organisms on which fish feed. Much of the acid rain debate revolves around whether these impacts on forests and waters are severe enough to warrant tighter regulation of emissions from coal-burning power plants.

Acid rain poses political problems because it often falls on areas far from the source of the sulfur dioxide emissions. The heaviest concentration of coal-fired power plants and sulfur dioxide emissions is in the Ohio Valley, particularly in Pennsylvania, Ohio, Indiana, and Illinois. The resulting acid rain, however, falls in upstate New York, New England, Canada, and other areas, carried by prevailing winds. It should be noted, however, that the most acidic rainfall occurs in places such as Kane, Pennsylvania, close to the sources of the sulfur dioxide.

For areas receiving the rainfall, the extent of environmental damage is only one side of the equation. Many parties feel they should not be forced to accept *any* pollution from another state involuntarily.

The U.S. effort on acid rain began June 30, 1980, when President Carter signed legislation (S. 932) that committed the government to a long-term study on acid rain. Barely one month later, on August 5, leaders from the United States and Canada signed a memorandum of intent to develop a formal treaty on acid rain control.

Canadian officials stepped up their criticism of the U.S. government during the Reagan administration, charging that the administration was not moving rapidly enough on the problem. Canadian Environment Minister Thomas McMillan charged in September 1985 that U.S. consumers should be willing to accept higher utility bills to pay for acid rain control. He noted that Canada generates only 5 million tons annually of sulfur dioxide, compared with 24 million tons in the United States.

McMillan then staked out a moral high ground by committing Canada to controls. "Before we go down to Washington and point a finger at the Americans, we have to exercise the moral authority of cleaning up ourselves and making sure our hands are clean," he said on September 15, 1985, at an acid rain conference. The Canadian program was to cut sulfur dioxide emissions in half.

McMillan added in January 1987 that the efforts of the United States to develop clean-coal technology, while useful for U.S. companies, were "largely irrelevant" to Canadian concerns.

One aspect of Canada's concern is the nature of its aquatic resources. A study published in February 1988 by Environment Canada stated that waters in 46 percent of Canada's land area—1.54 million square miles—are highly sensitive to acid rain. In Quebec, 82 percent of the land area is highly sensitive, according to Environment Canada. Another 21 percent of Canada's land was assessed as somewhat sensitive.

The Solution

The solution Congress devised in the Clean Air Amendments of 1990 was, in many ways, a political one. Legislators chose not to wait for the final report from the National Acid Precipitation Assessment Program, accurately surmising that its conclusions would not be so conclusive after all.

The final bill required 10-million-ton reduction in annual sulfur dioxide emissions and a 2-million-ton reduction in annual emissions of nitrogen oxides.

Phase I

Congress began by identifying the highest-polluting utilities. The bill singled out 110 plants having a total of 265 coal-fired boilers, all with a capacity greater than 100 megawatts and all in eastern and midwestern states. Table A in Section 404 of the law lists each utility unit.

In Phase I of the program, emission rates from these plants would be limited to 2.5 pounds of sulfur dioxide per million Btu of heat input, roughly double the standards for new plants of 1.2 ppm Btu.

Congress did not impose these requirements as absolute limits, however. Instead, it gave those utilities "allowances" that represented total emission limits—not emission rates—an essential distinction. The formula is simple: multiply the unit's average annual fuel consumption (expressed in Btu) by 2.5 ppm Btu. Each plant must have one allowance for each ton of sulfur dioxide emitted, beginning January 1, 1995.

Utilities that emit less than they are allowed may accumulate allowances for use in future years or may sell or trade them to other utilities. This is the key to Phase I: giving industry the flexibility to make the reductions where they are most cost-effective and using the marketplace to regulate the distribution of emision allowances.

Phase II

Five years after the start of Phase I, in 2000, Phase II takes effect. It will extend the allowance system to all U.S. utilities, and the sulfur dioxide emission rate on which the allowances are based will be tightened to 1.2 ppm Btu.

The law also will impose a total cap on utility emissions of 8.9 million tons annually, meaning that only 8.9 million allowances will be available. However, allowances accumulated during Phase I may be used to help a utility remain in compliance with the law.

Another provision that will help power plants meet the standards is a bonus allowance system, which will be in effect from 2000 through 2009. Each year, 530,000 allowances will be available.

Units in certain high-growth states will receive bonus allowances. Municipal power plants also are eligible. Of the annual bonus allowances, 50,000 will be distributed on a pro-rata basis among utilities in 10 states: Illinois, Indiana, Ohio, Georgia, Alabama, Missouri, Pennsylvania, West Virginia, Kentucky, and Tennessee. States with emission rates at or below 0.8 ppm Btu are eligible for bonuses as well.

EPA will create a reserve of allowances that it will sell in addition to those sold by utilities. Utilities that emit more sulfur dioxide than allowable must pay $2,000 per ton and must offset any excess emissions in the following year.

The acid rain program also will require reductions in emissions of nitrogen oxides; the goal is to reduce utility emissions by 2 million tons annually. The 1995 emission limit will be 0.45 ppm Btu or 0.50 ppm Btu, depending on the type of boiler. EPA also is directed to revise the new source performance standard.

President Bush and Prime Minister Mulroney signed a bilateral air quality accord in March 1991 that committed each country to reduced emissions of sulfur dioxide. The accord mirrored the emission reductions already required by the Air Act amendments.

Officials from the United States and Canada expressed hope after the signing that the accord would serve as a framework for efforts to control other pollutants. The sulfur dioxide emission limits are contained in an annex to the accord, and Canadian regulators already are working on a new annex that would address urban smog. It took 13 years of lobbying the U.S. government to obtain the sulfur dioxide commitments, and "we don't want to take 13 years on the next problem," Alex Manson, former director of acid rain policy for Environment Canada, said after the signing.

Air Toxics

Hazardous air pollutants are defined in Section 112 of the Clean Air Act as those "which may reasonably be anticipated to result in an increase in mortality or an increase in serious irreversible, or incapacitating reversible, illness." The standards themselves, referred to as National Emission Standards for Hazardous Air Pollutants (NESHAPs), are developed by EPA. Although environmentalists contend there are hundreds, perhaps thousands, of airborne substances that meet the Section 112 definition, EPA so far has regulated just eight: arsenic, asbestos, benzene, beryllium, mercury, radionuclides, radon-222, and vinyl chloride. (Actually, the number is seven if radon-222 is counted as a radionuclide.)

Instead of relying on individual SIPs to control hazardous air pollutant emissions, states develop implementation and enforcement procedures, which are subject to EPA approval. States have less flexibility in designing NESHAP enforcement programs than criteria pollutant control programs. NESHAPs apply nationwide.

The agency has not been able to develop more NESHAPs because the process is extraordinarily complex. The law stated that the EPA administrator was to publish a list of chemicals within 90 days of enactment of the law (in 1970). The agency then had 180 days to propose a NESHAP for each substance and another 180 days to publish final regulations. In reality, the process has taken much longer. For example, standards for benzene emissions from maleic anhydride and ethylbenzene/styrene plants were issued in final form on September 14, 1989—more than 12 years after EPA was given authority to develop NESHAPs by the Clean Air Act Amendments of 1977 (Public Law 95-95).

The need to develop a solid scientific basis for each NESHAP is one cause for delay. Epidemiological studies, which track actual human health effects data from exposed individuals, require extensive amounts of time and money to complete. EPA must make many tough decisions on the level of risk to allow, how many people are exposed to the substance, and how many incidences of cancer or other disease will be eliminated by various control options. Cancer is the principal disease of concern, although others, such as asbestosis, a debilitating lung ailment, are considered as well.

Once a body of data is collected, the calculation begins with a theoretical "most exposed individual." This would be someone who spends his or her entire lifetime (70 years) living near a pollution source. The risk to such individuals that the agency should tolerate is still subject to debate; however, for the benzene rule EPA split its risk calculation. For 99 pecent of the population, EPA set a risk level of 1 in 1 million. For the remaining 1 percent—presumably the "most exposed" portion—the risk was set at 1 in 10,000.

EPA adopted this dual risk approach because the risk to the "most exposed" population is still relatively small. For example, in its September 14, 1989, proposal to regulate benzene emissions from chemical manufacturing process vents, the agency said the exposure for that group would result in 0.01 cancer case per year—one every 100 years. To control emissions to reduce the risk below that level would be extremely costly and would have a negligible effect on human health. For the vast majority of the U.S. population (actually greater than 99.9 percent), the risk is far lower.

The agency still is grappling with the controversial risk-level issue, and some regulators are pushing for a uniform risk level or risk range for all media. The process also is subject to legal and political pressures, so the situation may change. However, EPA appears to have adopted the policy that while risks can be reduced, they usually cannot be eliminated entirely.

Another cause for delay, of course, is litigation. The key decision concerning NESHAP development came from the U.S. Court of Appeals for the District of Columbia Circuit in 1987 (*Natural Resources Defense Council v. EPA*, 824 F.2d 1146, 19 ER 294, 26 ER Cases 1263 (D.C. Cir. 1987)). In that case, Judge Antonin Scalia (later appointed as a Supreme Court justice) struck down a vinyl chloride NESHAP. He ruled that EPA could not develop a standard by simultaneously weighing human health risk and the cost and feasibility of the controls. Instead, Scalia said, EPA must apply a two-step process. First, it must set an acceptable risk level. Then and only then can EPA consider other factors such as cost, technical feasibility, and an adequate margin of safety.

The benzene rule was the first NESHAP developed under this court-ordered approach.

Hazardous Air Pollutants

Control requirements in NESHAPs vary according to the pollutants and their sources. For example, uranium mines can meet the requirements for controlling radon-222 emissions simply by sealing unused mine areas with airtight bulkheads. Beryllium sources must meet a 24-hour total emissions limit. Vinyl chloride plants are regulated according to the concentration of emissions in exhaust gas. Radionulcide emissions are regulated by millirems per year—a measure of how much radioactivity an individual would absorb from a source.

Regulations outlining NESHAP requirements are found in Section 40, Part 61, of the Code of Federal Regulations. Additional information on the chemicals can be found in BNA's *Chemical Safety Data Guide.*

Arsenic

This metallic element commonly is found in the compound arsenic trioxide, a white powder that itself is sometimes called arsenic, and arsenic trichloride. It is emitted during glass manufacturing, copper smelting, and arsenic production. This highly poisonous substance can be inhaled as dust and affect the skin, eyes, and respiratory system.

Existing glass manufacturers must limit total arsenic emissions to 2.5 metric tons per year. New plants must meet a standard of 0.4 metric ton per year. Primary copper manufacturers must place hood systems around smelters to capture arsenic emissions. Manufacturers of arsenic and arsenic trioxide must follow emission control procedures and monitor for visible emissions.

These standards apply to inorganic arsenic compounds only. Organic arsenic compounds such as arsanilic acid and cacodylic acid, which are used in pesticides, are not covered by this NESHAP.

Asbestos

This light, fibrous element once was widely used in insulation, fire-proofing, and other construction materials. Because its tiny fibers can flake off and become suspended in the air, they can be inhaled. When inhaled, some asbestos fibers can penetrate deeply into the lungs and lodge there permanently. They can cause a fatal lung illness known as asbestosis, and also can cause lung cancer.

The general standard for asbestos mills and disposal sites is that they must not generate visible emissions of asbestos. Alternatively, they can satisfy the NESHAP requirements by following EPA-approved emission control procedures. For companies demolishing buildings that contain asbestos, all removable asbestos should be taken from the building before demolition. If it is encased in concrete or otherwise protected, the asbestos must be kept wet during demolition. The resulting waste must be taken to an adequate asbestos disposal site.

Benzene

This colorless, flammable liquid evaporates at room temperature and is easily inhaled. Its main industrial source is steel industry coke ovens. It also is found in gasoline.

Benzene can be acutely hazardous if inhaled, causing headaches, nausea, drowsiness, unconsciousness, and other symptoms. It can irritate the eyes, throat, and nasal passages. It also can damage the brain, liver and kidney. Finally, tests indicate that it may cause leukemia.

Coke production involves heating coal in an oxygen-starved atmosphere to drive off volatile organic compounds; the product is used in iron and steel production. The by-product, which includes benzene and other volatile organic compounds, is drawn off and condensed.

Coke oven by-product facilities have standards of their own. The traditional control measure for most facilities is enclosing equipment that houses or transports benzene. Facility owners also must control *fugitive emissions,* defined as leaks from pumps, valves, connectors, and other pieces of equipment.

Beryllium

This rare, light, metallic element is generally used in production of metal alloys and in rocket fuel. It is emitted by plants processing the ore, foundries and machine shops that use beryllium alloys, and rocket motor testing.

Beryllium is a known animal carcinogen and a suspected human carcinogen. Inhalation of beryllium can cause an illness known as chronic beryllium intoxication, and it can cause inflammation of the eyes and nasal passages.

All sources except rocket motor testing must meet a standard of 10 grams of total emissions over a 24-hour period. Rocket motor testing facilities must prevent ambient concentrations from exceeding hazardous levels, measured using a complex time-weighted formula.

Mercury

This heavy metallic element is emitted in mercury ore extraction and processing, from mercury chlor-alkali cells used in chlorine gas production, and from sewage treatment sludge incineration. When inhaled, it attacks the human nervous system and can cause multiple chronic and acute disorders, such as headaches, nausea, chest pains, diarrhea, rashes, loose teeth, and neurological symptoms.

The standard for ore processing and chlor-alkali cell emissions is 2,300 grams of mercury over a 24-hour period. Sludge incinerators must meet a limit of 3,200 grams over a 24-hour period.

Radionuclides

A radionuclide is not a substance but a class of substances, defined as any radioactive atom. For EPA's purposes, a radionuclide may be defined as any radioactive substance that can be inhaled.

Sources of radionuclides include power plants, nuclear weapons facilities, hospitals and other medical facilities, uranium mill tailing piles, and phosphorus plants. Lung cancer is the principal health problem related to radionuclide exposure.

Nuclear weapons plants must meet an emission standard of 10 millirem per year (effective dose equivalent).

Many of the radionuclide sources covered by NESHAP also are licensed by the Nuclear Regulatory Commision (NRC). Some facility operators and their allies in Congress have complained about dual regulation, charging that EPA's more stringent emission limits are unnecessary and burdensome. EPA responded by staying the standards while it studies whether the NRC rules provide adequate protection of human health.

Radon-222

This radioactive element (222 is its atomic weight) occurs naturally, and also is released by uranium mining. It can cause cancer when inhaled.

As mentioned above, mine operators can meet the standard by sealing unused or abandoned parts of their mines. They also can apply to EPA for alternative standards. Only underground mines are covered by this NESHAP.

Vinyl Chloride

This colorless, flammable gas is used and emitted during production of plastics. Some plants manufacture it, and others generate it as a by-product. Vinyl chloride is a known human carcinogen. It also can cause severe eye irritation, nausea, and intoxication.

Manufacturers of vinyl chloride, ethylene dichloride, or polymers (large molecules made up of many smaller molecules) containing fractions of polymerized vinyl chloride cannot emit vinyl chloride in concentrations exceeding 10 parts per million, averaged over a three-hour period. The NESHAP also restricts emissions from vents and sources of "fugitive emissions," such as equipment leaks.

The New Air Toxics Regime

Congress expressed profound dissatisfaction with EPA's regulation of air toxics during the debate on the Clean Air Amendments of 1990. In response, the lawmakers imposed on the agency a requirement to regulate 189 hazardous air pollutants. They are listed by name in the legislation.

The bill also informs EPA as to what facilities to regulate and how to develop the standards. Any source that emits more than 10 tons annually of a listed chemical or 25 tons annually of any combination of listed chemicals is subject to regulation.

Health-based standards for hazardous air pollutants generally depend on epidemiological studies, which track the actual effects of chemical exposure on humans and animals. Such studies are imprecise and difficult to control scientifically, but the main problem is that they require an extensive amount of time. Legislators and environmentalists pushed for a different basis: existing technology.

The new air law directs EPA to assess the current control technologies for each of the listed pollutants and set a "maximum achievable control technology" standard for each pollutant source. This is based on the average emission reduction of the most effective 12 percent of the sources, or the best 5 sources in categories with fewer than 30 sources. Such standards may accommodate cost, energy demands, and other effects on human health and the environment.

Six years after enactment, EPA is required to evaluate the effectiveness of the technology-based standards, in consultation with the U.S. surgeon general. Eight years after enactment, the agency must publish new standards for any source if it considers such standards necessary to protect human health. The standards must provide an ample margin of safety to protect public health unless the agency determines an even more stringent standard is needed to protect the environment.

The public may petition EPA to add chemicals to or remove them from the list.

Industry representatives voiced their reservations about the new air toxics program. According to a February 1990 study by the CONSAD Research Corp., more chemical plants (at least 250) will be affected than any other manufacturers, followed by paper products, lumber and wood products, petroleum refining, primary and fabricated metals, transportation equipment, stone/clay/concrete, and rubber/plastics.

The impact on jobs will be quite different. More trnsportation equipment jobs will be affected (more than 200,000) than any other industry, followed by primary metals, chemicals, paper products, electronics, petroleum refining, and industrial machinery, according to the CONSAD report.

The electricity and gas supply industry will be affected most among the nonmanufacturing industries. The oil and gas extraction, nondurable goods, national security, coal mining, and health service industries also will feel an impact.

Jobs will be created in the pollution control industry. The Congressional Research Service estimated, in January 1990, that the acid rain control program is likely to generate 50,000 jobs through manufacture of scrubbers to control sulfur dioxide emissions from power plants. The money spent by business to comply with the air toxics program certainly will enrich the pollution control industry; whether this will offset the losses remains to be seen, however.

PERMITS

Permits are required for any major source of pollutants covered by NAAQS or NESHAP. The permitting authority is either EPA or a state with an EPA-approved permit program.

The permit program was created by the Clean Air Act Amendments of 1990. Some states and cities already had permit programs in place; for many plants, however, the permit requirements were completely new.

The Air Act requires permitting authorities to write emission limits, compliance schedules, and monitoring requirements into each permit. Violations of permit conditions are enforceable by the permitting authority, which can recover civil penalties in court up to $10,000 per day of violation.

The permit program imposed a significant new burden on manufacturers, but it also provided a new protection: factories that are in compliance with the conditions of a valid permit will be shielded from enforcement action. Because the program is relatively new, the breadth and effectiveness of this shield has not been tested in court.

ENFORCEMENT

As in other environmental laws, the EPA administrator has authority under the Air Act to issue orders, assess civil penalties, and file civil and criminal actions in federal court. All court action on EPA's behalf is conducted by the Department of Justice.

The agency's authority varies by the type of violation. For violations of SIPs, the administrator first must give the violator and state 30 days' notice. If compliance is not obtained thereafter, the administrator may order the party to cease the violations, assess a penalty up to $25,000 per day for each violation, or file a civil action in federal district court.

Orders by the administrator cannot take effect until the alleged violator has an opportunity to confer with EPA on the violations. An order must result in compliance "as expeditiously as practicable," but in no case more than one year from the date of the order.

Administrative penalties generally apply to violations of SIPs; titles I, III, IV, V, and VI of the law; and new source requirements. If the total amount sought in penalties exceeds $200,000, the administrator must press the case in federal district court. Alleged violators have an opportunity for a public hearing, and also may challenge a penalty in federal court.

The administrator may compromise on the final amount of penalties, and such compromises are common. In fact, final penalties often represent only a fraction of the original penalty assessment. Any settlement or consent order is published in the *Federal Register* and is subject to 30 days' public comment.

A state failure to enforce its own SIP can result in EPA assuming enforcement authority. A recent example of such action came in October 1990, when EPA proposed a federal implementation plan to control carbon monoxide in Phoenix and Tucson, Arizona. A federal appeals court ruled that EPA should not have approved Arizona's plan because it contained only three of the 12 control measures recommended in an agency study (*Delaney v. EPA*, 898 F.2d 687, 31 ER Cases 1001 (9th Cir. 1990)). The court then ordered EPA to propose its own plan in six months.

The proposed plan was never made final, however, because the Clean Air Act Amendments of 1990 (signed in November) changed the requirements for cities' compliance. The agency's capability to assume enforcement powers remains, however, and now extends to failures to enforce the new emission permit program as well.

If a state fails to enforce requirements for construction of modification of major air pollution sources, the administrator may prohibit construction of any new source in the affected area.

For violations of the Air Act that are unrelated to SIPs, the administrator's authority is similar. The major difference is that knowing violations are considered criminal and are punishable not only by orders and fines, but by jail sentences.

A party who knowingly violates an SIP, an EPA order, the Air Act, or its regulations is subject to fines under Title 18 of the U.S. Code (which covers all federal criminal action) and up to five years in jail. One who submits false statements to EPA, fails to submit required information, or tampers with monitoring equipment is subject to fines and up to two years in jail. Failure to pay fees is punishable by fines and up to one year in jail.

Negligent releases of hazardous air pollutants also are subject to fines and jail terms up to one year. A party who releases a hazardous air pollutant and knows or should know that it creates "imminent danger of death or serious bodily injury" can be punished by fines and up to 15 years in jail. Criminal penalties may be doubled for repeat offenders.

Citizens are empowered to sue private parties or government authorities (including the U.S. government) for violation of any emission limit or order and for new source infractions. Citizens also may sue EPA for failing to enforce the law. They must give 60 days' notice before filing suit in federal district court.

EPA is empowered to "intervene as a matter of right at any time in the proceeding" of a citizen enforcement action. A citizen suit may not proceed if EPA can show that government authorities are pursuing diligent prosecution in state or federal courts.

A new provision of the law establishes a special penalty fund in the U.S. Treasury, financed by fines for Air Act violations. It will be used by EPA to cover enforcement expenses.

SUMMARY

The following are some important points to remember about air pollution programs:

EPA sets NAAQS for criteria pollutants; these standards are expressed in parts per million, except for lead and particulates, expressed in grams per cubic meter.

Standards for hazardous air pollutants, also set by EPA, vary according to the nature of the pollutant. Some impose an emission rate or a total emission limit. Others require installation of certain control equipment. Still others specify workplace practices.

States develop SIPs for areas that exceed one or more NAAQS. All SIPs and all SIP revisions must be approved by EPA.

NSPS, issued for specific industrial categories, apply to all new plants and all plants that undergo significant modifications. These standards can take many forms.

Existing plants are never covered by an NSPS, but can be regulated by states in the SIP process. Industries wishing to build new plants in areas that meet NAAQS must comply with the PSD program.

II. CLEAN WATER ACT

INTRODUCTION

Federal regulation of water quality has the longest history of any pollution control program, beginning in 1948 with the first federal aid to municipalities to construct sewage treatment plants. The modern era of water pollution control really began, however, with the federal Water Pollution Control Act Amendments of 1972 (Public Law 92-500), one of several environmental statutes enacted in the wake of Earth Day in 1970.

The bill revamped the grants system for sewage treatment projects; perhaps more important, however, it established a national permit system for discharges into U.S. waters. The legislation also put industries on a schedule to install best practicable control technology by 1977 and best available control technology by 1983. It established requirements for effluent limitations, water quality standards, and industrial pretreatment as well.

The most recent major change in legislation came with the Water Quality Act of 1987, unanimously passed by Congress in 1986 and vetoed by President Reagan. On February 4, 1987, the 100th Congress enacted the bill (Public Law 100-4), over Reagan's veto, as one of its first acts.

This bill began the phase out of federal aid for sewage treatment plant construction. It also created a new system to regulate stormwater discharges. Another section required states to identify waters with toxic contamination problems, locate the sources of contamination, and impose appropriate controls. It left in place, however, key elements of the 1972 law, including the national permit system.

THE PERMIT SYSTEM

The foundation of the Clean Water Act's protection of navigable surface waters is the discharge permit. Every entity that discharges pollutants into navigable waters is required to have a permit, whether it is municipal, industrial, or individual. To control these discharges, the Environmental Protection Agency (EPA) has developed the National Pollutant Discharge Elimination System (NPDES), and the permits issued under the system are known as NPDES permits.

The two principal types of discharges are domestic sewage and industrial wastewater. Municipal sewage treatment plants all require NPDES permits and are dealt with under a specific EPA "national municipal policy." Industries are divided into categories and given discharge limits.

Some industries discharge directly into U.S. waters, and must obtain NPDES permits from the state or EPA. Other industries discharge to municipal sewer systems, and must adhere to any "pretreatment" requirements set by the municipal treatment authority.

What Are Navigable Waters?

Navigable waters, for the purpose of Clean Water Act permitting, are broadly defined. Section 40, Part 110.1, of the Code of Federal Regulations defines navigable waters as:

(a) Waters that are used or may be used in interstate commerce (i.e., transportation, fishing, recreation, etc.);

(b) Interstate waters, including interstate wetlands;

(c) Lakes, rivers, streams (including intermittent streams), mudflats, sandflats, and wetlands;

(d) Any "impoundments" (enclosure or partitioning, such as for irrigation) of navigable waters; and

(e) Tributaries of any of the above waters, including adjacent wetlands.

"Navigable waters" and "U.S. waters" are virtually interchangeable terms for the purpose of Clean Water Act permitting, standards, and enforcement.

The U.S. Supreme Court issued a pivotal decision in 1985 to help define the waters covered by Clean Water Act permit programs (*U.S. v. Riverside Bayview Homes*, 106 S. Ct. 455, 23 ER Cases 1561, Dec. 4, 1985). A developer in Hamilton County, Michigan, sought to fill a marshy area adjacent to open water and build homes on the land. The U.S. Army Corps of Engineers has jurisdiction over permits for discharges of dredged or fill material into navigable waters, and asserted that a permit was needed to fill the area. The developer countered that the area had been flooded only five times in 80 years and was in no way navigable.

The Supreme Court unanimously upheld the authority of the Corps of Engineers. The court's first finding was that "any discharge of dredged or fill material into 'navigable waters'—defined as 'waters of the United States'—is forbidden unless authorized by a permit issued by the Corps of Engineers" under Section 404 of the act.

The court then determined that although the lands were flooded infrequently, a connection to navigable waters through groundwater certainly existed; the land was swampy because of stable hydrologic conditions, not occasional flooding. It also supported typical wetlands vegetation, the court noted.

Finally, the high court held, Congress intended the Clean Water Act to give the corps broad jurisdiction to regulate discharges into wetlands.

Water Quality

The Clean Water Act states that "effluent limitations" are needed to control any discharges that "would interfere with the attainment or maintenance of that water quality in a specific portion of the navigable waters which shall assure protection of public health, public water supplies, agricultural and industrial uses, and the protection and propagation of a balanced population of shellfish, fish and wildlife, and allow recreational activities in and on the water."

States must determine what uses are appropriate for water bodies; only then can decisions be made on how to protect human health and aquatic life. The three general use standards cover fishing, swimming, and drinking, with the standards for a drinking water supply being the most stringent. The "fishable" standard is the least stringent, but, even so, protection of aquatic life can require tight effluent limitations.

Every state has gone through the exercise of designating waters for various uses. In March 1972, for example, the Illinois Pollution Control Board approved regulations specifying that all state waters except the Chicago River and the Little Calumet River were designated for public and food processing water supply use. At the same time, the board gave communities and industries until December 31, 1973, to install treatment facilities to remove 99 percent of the waste from sewage.

The 1987 Clean Water Act amendments required states to determine what water bodies are not meeting their designated uses because of toxic pollution and to develop plans to control the toxic discharges responsible (see later section on state toxics programs).

Effluent Limitations

The numeric limit on pollutant discharges, usually expressed in grams per liter or parts per million (ppm), is known as an "effluent limitation." Part 40, Section 401, of the Code of Federal Regulations lists 65 toxic pollutants for which EPA or states must issue effluent limitations, from acenaphthene to zinc. It also requires standards for five conventional pollutants: biochemical oxygen demand (BOD); total suspended solids (TSS); acidity, expressed as pH; fecal coliform; and oil and grease. All but acidity and fecal coliform are measured as mass per volume of water, usually grams per liter.

Here a distinction should be made between effluent limitations, which are the actual standards issued by EPA or states, and effluent limitations guidelines, which are the regulations published by EPA to guide development of those standards. The guidelines contain a recommended standard for each pollutant, and states must justify any less stringent standards.

The most recent example of states developing less stringent effluent limits than recommended by EPA involves dioxin. In September 1990, the agency approved Maryland's standard of 1.2 parts per quadrillion in pulp and paper mill wastewater. EPA's recommended limit is 0.013

part per quadrillion. Maryland convinced regulators that use of the less stringent Food and Drug Administration cancer risk estimate was appropriate.

BOD is a measure of a pollutant's capacity to deplete oxygen in a water body. It does so by providing nutrients that promote the growth of microscopic organisms that consume oxygen. Fish and other aquatic organisms can die as a result. Human and animal wastes, detergents, and fertilizers are common sources of BOD. The prevailing method for expressing BOD, known as BOD5, involves multiplying the levels of fats, proteins, and carbohydrates by factors of 0.890, 1.031, and 0.691, respectively. Some industries also must control for a related pollutant, chemical oxygen demand (COD).

TSS refers to any of the solids that cannot be readily filtered from wastewater. Suspended solids are generated throughout U.S. industry, and virtually every set of effluent limits includes standards for suspended solids. Certain industries are allowed to discharge higher levels of solids than others, however. For example, seafood processors typically must meet standards in the range of kilograms per 1,000 kilograms; best conventional technology for shucked oyster processing limits discharges of suspended solids to 45 kilograms per 1,000 kilograms. On the other hand, most limits on suspended solids are expressed in milligrams per kilogram—1,000 times more stringent.

Acidity is a measure of the positively charged hydrogen ions in an aqueous solution. Neutral acidity is considered 0.0000001 gram of hydrogen ion per liter (10-7), or pH 7. Because the pH measure is based on a negative exponent (in this case -7), a lower number means *higher* acidity. Thus, a pH of 3 would be highly *acidic*; a pH of 12 would be highly *basic*, with a surfeit of negative ions. Another feature of this scale, known as a logarithmic system, is that differences of one unit actually indicate tenfold differences. Thus, a solution with a pH of 2 is 10 times more acidic than a solution with a pH of 3.

Fecal coliform bacteria is found in domestic sewage and farm animal wastes. Fecal coliform levels generally are expressed as a number of organisms per 100 milliliters (i.e., 20/100 milliliters). Often, total coliform is measured instead; water quality and drinking water standards tolerate higher levels of total coliform than fecal coliform. Some tests for fecal coliform contamination are simply positive-or-negative tests, and the related standard would set a maximum percentage of positive readings.

Oil and grease generally means those substances in any form. Section 301 of the act defines oil, for example, to include petroleum, fuel oil, sludge, oil refuse, and oil mixed with wastes other than dredged spoil.

INDUSTRIAL SEWAGE TREATMENT

The Clean Water Act sets different standards for sewage treatment plants than it does for other industrial categories, but all dischargers require NPDES permits and all must meet limits for both conventional and toxic pollutants.

Section 306 of the Clean Water Act requires EPA to publish effluent limitations for the following industries:

- Pulp and paper mills
- Paperboard, builders' paper and board mills
- Meat product and rendering processing
- Dairy product processing
- Grain mills
- Canned and preserved fruits and vegetables processing
- Canned and preserved seafood processing
- Sugar processing
- Textile mills
- Cement manufacturing
- Feedlots
- Electroplating

- Organic chemicals manufacturing
- Inorganic chemicals manufacturing
- Plastic and synthetic materials manufacturing
- Soap and detergent manufacturing
- Fertilizer manufacturing
- Petroleum refining
- Iron and steel manufacturing
- Nonferrous metals manufacturing
- Phosphate manufacturing
- Steam electric power plants
- Ferroalloy manufacturing
- Leather tanning and finishing
- Glass and asbestos manufacturing
- Rubber processing
- Timber products processing

The above industries are required to meet effluent limits for any conventional and toxic pollutants they discharge to navigable U.S. waters.

In addition, all industries that discharge directly to U.S. waters must file discharge monitoring reports that document the daily level of pollutants released to those waters. This requirement, incidentally, allows environmental and citizens groups to check industry compliance with effluent limits, because the discharge monitoring reports are open to the public.

Industries still are struggling to reach complete compliance with the law. Based on information supplied by municipal sewage treatment plants, 41 percent of all industrial discharges violated at least one standard over a one-year period, according to a May 1989 General Accounting Office (GAO) report. Self-reporting by industries indicated a lower—but still significant—violation rate of 20 percent, GAO said.

These violations can have serious consequences for surface waters if sewage is discharged directly, or for treatment plants if it flows through the sewer system. According to GAO, in one case a sewage treatment plant worker was killed by toxic fumes. Another company's violation effectively disabled a treatment plant for months. In a third case, toxic chemicals in sewage sludge forced a municipality to scale back use of a sludge landfill.

BCT, BPT, and BAT

Section 301 of the Clean Water Act directs EPA to write effluent limitations for industrial wastewater discharges that require *best conventional pollution control technology* (BCT), *best practicable control technology currently available* (BPT), or *best available technology economically achievable* (BAT), depending on the pollution and source involved. EPA defines these terms according to a list of items regulators must consider when imposing standards. Any standard must accommodate the age of the equipment, the process, engineering aspects, process changes, and other environmental impacts. Cost also must be considered, although it is handled differently for each type of standard.

Standards based on BCT are the least stringent of the three. In general, there must be a reasonable relationship between the cost of the controls and the benefits. The standard writer also must compare the cost to industry of controlling a given pollutant with the cost to municipal sewage treatment plants.

Standards based on BPT also must account for costs and benefits. They do not involve, however, any comparison with costs to sewage treatment plants.

Standards based on BAT are the most stringent. Costs may be considered when writing the standards, but EPA must weigh them only as they affect specific industries.

They may not be weighed against the benefits of reducing or eliminating pollutant discharges. (Definitions from 40 CFR 125.2)

Sections 301 and 307 specify certain chemicals whose discharge is subject to BAT standards.

MUNICIPAL SEWAGE TREATMENT

Regulation of municipal sewage treatment discharges also is based on a three-tiered regime: primary, secondary, and advanced treatment. Unlike industrial treatment, however, each tier represents a type of treatment, not an increasingly stringent treatment standard. Primary treatment consists of removing all solids so that the treated water is clear, although it still may contain many pollutants. Secondary treatment involves removal of biological nutrients and organisms that feed on those nutrients. Advanced treatment involves removal of toxic pollutants. As a result of EPA requirements, virtually every municipal sewage treatment plant provides both primary and secondary treatment.

Primary treatment begins with large gates that prevent solid objects (shoes, clothing, wood, and plastic bottles) from entering the system. Then the sewage goes to settling ponds where the solids gradually sink to the bottom or float to the top.

Secondary treatment is far more complicated. Nutrient matter must be removed from domestic wastewater before being discharged; otherwise, the nutrients would overenrich the receiving waters. Such enrichment causes algae to grow, shutting off sunlight to bottom-dwelling plants and robbing the water of oxygen. Both fish and plant life may vanish as a result. Another benefit of secondary treatment is that it helps control waterborne diseases.

Secondary treatment consists of exposing domestic sewage to certain bacteria that consume the nutrients. These bacteria die when the nutrients are consumed and settle out of the treated water.

There are two measures of the effectiveness of secondary treatment. One is the level of fecal coliform bacteria, a class of microorganism generally found in domestic sewage. The other is BOD, or the pollutants' capacity to deplete oxygen in receiving waters.

Advanced treatment usually is necessary only when industries discharge toxic pollutants into a sewer system. Such discharges require the permission of treatment plant officials. The treatment provided by the plant would vary according to the nature of the toxic pollutant.

PRETREATMENT PROGRAMS

If a company discharges wastewater directly into navigable waters, it is required to have an NPDES permit; if it discharges into a public sewer system, it is not required to have such a permit. However, the Clean Water Act allows the local sewage treatment authority to impose conditions on industrial discharges within its system. This is known as the pretreatment program.

Section 307(c) of the Clean Water Act authorizes the EPA administrator to issue pretreatment standards for any industrial category for which the administrator also has issued effluent limitations. It further states: "Such pretreatment standards shall prevent the discharge of any pollutant into such treatment works, which pollutant may interfere with, pass through, or otherwise be incompatible with such works."

A typical example of interference is the discharge of a toxic chemical that, when it reaches the secondary treatment unit, kills the bacteria that make the system work, resulting in ineffective removal of nutrients. "Pass through" simply refers to a toxic chemical that would flow through the treatment process without being removed. It involves chemicals the plant is not designed to treat. The pretreatment program also prohibits the discharge of any substance that would endanger the health or safety of sewage treatment plant workers.

An industrial discharger may need to comply with a federal pretreatment standard, or it may be subject to a state or local standard. Such pretreatment can take a variety of forms. An early example surfaced in Sumter, South Carolina, where Santee Print Works spent $100,000 in 1972 on a treatment facility to remove floatable solids and print wastes from its sewage. This enabled the local sewage treatment plant—under construction at the same time—to treat the plant's wastewater.

On a more ambitious level, also in 1972, International Harvester spent $3.6 million on a treatment plant to process 36 million gallons of water per day at its Wisconsin Steel Works in Chicago. The effort was part of a drive to close outfalls to the Calumet River.

More recently, EPA has been pressing both cities and industries to improve their pretreatment program compliance. In 1989, EPA brought 61 enforcment actions against cities for pretreatment violations. In October 1989, the agency sued Detroit, Phoenix, Arizona, and El Paso and San Antonio, Texas, for failing to control industrial discharges to their treatment plants. EPA Administrator William K. Reilly announced then that as many as one-third of the 1,500 cities with pretreatment programs may be out of compliance. Attorney General Richard Thornburgh called the pretreatment program "the weak link for removal of toxic pollution from our nation's waterways."

The federal government also has acted against individual dischargers. In February 1990, EPA obtained a consent decree with Toledo Coke Corp. under which the company agreed to pay a $100,000 fine and construct a treatment system to prevent future violations of pretreatment standards (*U.S. v. Toledo Coke Corp.*, No. 87-0117218, (N.D. Ohio, Feb. 5, 1990)).

STATE TOXICS PROGRAMS

The above permit-based programs have resulted in great improvements in water quality, but some persistent problems remain. Thus, Congress established, in the Water

Quality Act of 1987, a new state-based program to address persistent toxic pollution. The program also applies to "thermal discharges" because changes in water temperature can have profound effects on fish populations.

Section 303(d) [1313] of the law requires every state to identify all waters for which effluent limitations adopted under Section 301(b) are insufficient to bring about compliance with any applicable water quality standard. These waters are ranked by severity of contamination. States then must calculate the maximum daily loading of pollutants to those waters necessary to meet water quality standards.

Once the identification process is complete, states must adopt areawide waste treatment management plans under Section 208 [1288] of the law. In this process, states are to identify the sources of the toxic pollution and impose controls necessary for compliance with water quality standards.

By April 1990, 47 states had imposed 586 fishing advisories and 135 fishing bans as a result of their assessments of contaminated waters. About one-third of the monitored river miles have elevated levels of at least one toxic pollutant. Most of the contamination, EPA said in a status report, came from industrial discharges and releases from landfills. States currently are developing facility-specific discharge limits for the industries that generated the pollutants.

WETLANDS

The term "navigable waters" is extremely broad and encompasses wetlands, better known as swamps. Far from being the nuisances they were considered in the past, we know today that wetlands serve many purposes, including control of flooding and pollution. Wetlands are important spawning areas for fish and shellfish, both freshwater and saltwater, and are important nesting areas for migratory birds. Many species also make their permanent homes in wetlands.

Discharges to wetlands require an NPDES permit, just as any other discharges. The most common type of "discharge" is dumping of fill material (usually dirt or other clean fill) to convert the wetlands to dry lands. This is often done to expand agricultural property or for residential or commercial development.

Dredging, which converts wetlands to open water, similarly requires a permit. Both types of dredging and filling are covered by Section 404 of the Clean Water Act. The principal regulator, however, is not EPA but the U.S. Army Corps of Engineers (Corps), which has primary authority over maintenance of navigable U.S. waters. EPA has veto authority over those permits, however.

EPA and the Corps seldom issue Section 404 permits for wetlands conversion unless there is no alternative. Ports and marinas, for example, rely on their proximity to water bodies. Airports are often sited adjacent to water bodies for noise and safety reasons. Necessary expansion of such facilities would be eligible for Section 404 permits.

Basic commercial development, however, is usually not eligible. The notorious Attleboro Mall dispute in 1985 is a case in point. The developers sought to fill parts of Sweeden's Swamp in Massachusetts. They told regulators they had no alternative to filling the wetlands because they had already purchased the land. EPA rejected their argument, ruling that a shopping mall does not depend on proximity to a water body.

More recently, EPA went a step further and denied permits to the Two Forks dam and reservoir project in Colorado. In November 1990, the agency told suburban Denver water suppliers that the project would destroy irreplaceable wetlands on the South Platte River, as well as valuable wildlife habitats and trout fisheries. Moreover, the applicants failed to demonstrate that they had considered alternatives, in this case water conservation programs.

With water supply a key issue in the western United States, EPA sent an important message to cities: wetlands will not be sacrificed in the name of development.

NONPOINT SOURCE POLLUTION

So far, our discussion has focused on *point source* discharges into U.S. waters. They are readily identifiable and usually come out of a pipe or container from a single entity. But what about *nonpoint source* discharges? These are generally defined as polluted waters that begin as rainwater but pick up contaminants as they wash across city streets, parking lots, construction sites, farms, animal feed lots, and other surfaces. EPA officials estimate that up to half of all U.S. water pollution now is caused by nonpoint sources.

Sometimes this polluted rainwater (often called *stormwater*) flows directly into U.S. waters. In other cases, especially in urban areas, it flows into sewer systems. If this creates problems, the local sewage authority is responsible for solving them.

The stormwater permit provisions in the Clean Water Act resulted from a lawsuit pressed by the Natural Resources Defense Council. In 1975, a U.S. district court ruled that every municipal stormwater *outfall* (a pipe or other conduit that empties into surface waters) must have a permit (*NRDC v. EPA*, 396 F. Supp. 1386, 7 ER Cases 1881 (D.D.C. 1975)). This threatened a major change for towns and cities, and Congress intervened. Another key player was the Association of State and Interstate Water Pollution Control Administrators.

When Congress pieced together the Water Quality Act of 1987 in October 1986, it agreed to a new section on stormwater permits. Any municipal sewer system serving more than 250,000 people would need a permit, and EPA could require permits for smaller entities if their stormwater was causing pollution problems. The law drops the threshold to 100,000 residents after October 1992. This is expected to cover about 225 cities and counties.

The law also required an estimated 50,000 industrial sites to obtain permits. EPA has developed three separate permit systems, *individual*, *group*, and *general*.

Any stormwater discharger that is required to obtain a permit may satisfy the law with an individual permit. This traditional NPDES permit would be reviewed much like any other. It could include pollutant discharge limits and monitoring and reporting requirements.

Dischargers are eligible for group permits if they fall into the same effluent guideline category or their discharges are of similar type and volume. Part 1 of the permit application must list all participants along with their locations and information on their industrial activities. Applicants identify 10 percent of the group, which will submit Part 2 permits with much more detailed quantitative information. The Part 2 group must have a minimum of 10 members, but need not exceed 100 members if the entire group numbers more than 1,000.

Finally, many industries will be eligible under the general permit program. Under this program, no permit application need be filed, and no permit would be issued. A facility would simply monitor its discharges, keep records, and meet certain restrictions. The information must be sent to EPA if the agency requests it; otherwise, however, a facility operator must only inform EPA that it intends to participate in the general permit program.

The general permit would cover states that do not yet have authority to run the NPDES permit program and states that have partial authority but lack general permit program powers. Agency officials have said they expect most states to adopt similar programs, however, because they would ease the administrative burden of the new stormwater requirments.

The general permit would prohibit any discharges of hazardous substances in excess of the reportable quantities limits set for those substances under the Resource Conservation and Recovery Act. It also would ban any discharges in storm sewers that do not constitute stormwater—in essence, illegal hookups to storm sewers.

Those covered by general permits would need to develop pollution prevention programs. According to EPA, such programs might include, for example, covering all road salt storage areas to eliminate contact with rainfall and erosion control programs at construction sites.

The general permit may also include specific effluent limitations. These limitations would cover coal piles, landfills, oil and gas development, and other industries. Testing of the whole effluent for acute toxicity would be required for some industries.

Chemical manufacturing, construction sites, hazardous waste management, recycling, mining, and the lumber industry are expected to be significantly affected by the rules. EPA officials have said they hope the program will focus more on prevention of pollution than on end-of-pipe solutions. This would point toward practices such as better storage of materials to prevent contact with rainfall or tighter control of leaks and spills.

Most facilities are expected to be eligible for the general permit program. Exceptions would be facilities already subject to stormwater standards in their NPDES permits or in effluent limitations guidelines, facilities that EPA considers likely to cause water quality problems, and inactive mineral development sites on federal land.

For cities, the problem is different. When determining what cities should do to address water quality problems caused by stormwater discharges, EPA officials say they will look to what other cities have done. Possibilities include such practices as improved street cleaning to limit oil and grease in runoff and retention basins to allow solids to settle before reaching surface waters.

A city's course of action also will depend in part on what type of sewer system it has. Here we must make a distinction between *combined sewer systems* and *separate sewer systems*. Combined systems channel all wastewater—domestic sewage, industrial sewage, and stormwater—into one set of sewer lines that flow to a sewage treatment plant. Separate systems channel domestic and industrial sewage to treatment plants, while the stormwater is discharged directly, without any treatment, to U.S. waters.

Older, eastern U.S. cities are most likely to have combined water systems, while newer cities and suburbs generally are served by the more modern separate systems.

Separate sewer systems have the advantage of sending lesser volumes of wastewater to treatment plants. They also avoid the problem of combined sewer overflows: during heavy rains sewers can back up, and when combined systems back up, the overflow is composed of rainwater mixed with sewage. This untreated sewage may then run directly into U.S. waters or cause other localized problems.

Federal, state, and local governments have yet to come up with a solution to the combined sewer overflow problem. Because replacing a city's entire sewer system entails enormous cost, people are exploring such alternatives as large retention basins.

ENFORCEMENT

Section 309 of the Clean Water Act gives EPA broad enforcement authority. The agency notifies a state whenever a violation comes to its attention, and if that state fails to take appropriate action within 30 days, EPA is empowered to commence enforcement action itself.

Such violations could be actions that are inconsistent with provisions of the law itself or inconsistent with conditions of permits issued by states under the act, and they include discharges to U.S. waters and discharges to sewer systems. This enforcement authority applies to violations of effluent limitations and national performance standards (sections 301, 302, and 306); pretreatment standards (Section 307); recordkeeping, reporting, and inspection requirements (Section 308); special aquaculture provisions (Section 318); NPDES permit requirements (Section 402); standards for discharge of dredge and fill material (Section 404); and controls on sewage sludge disposal (Section 405).

The agency has *administrative* authority to issue compliance orders and levy civil penalties for alleged violators of any of the above sections and any permit requirements. EPA may, in tandem with the U.S. Department of Justice, file a *civil* action in federal district court for appropriate relief, including temporary or permanent injunctions. The law also established three classes of *criminal* violations: negligent, knowing, and knowing paired with imminent danger of death or serious bodily injury.

The agency's administrative compliance orders are designed simply to halt the violations. The same violations may give rise to administrative penalties as well. The administrator is empowered to levy Class I penalties up to $10,000 per violation or $25,000 total, or Class II penalities up to $10,000 per day with a $125,000 maximum. Alleged violators are entitled to a hearing before the EPA administrative law judge.

Note that all the above fines are administrative penalties. The law also sets limits on civil penalties that EPA may seek in court. Parties that violate the law or permit conditions are subject to court-imposed civil penalties of up to $25,000 per day of violation. If the violator is a municipality, the state will be brought in as a defendant as well.

Criminal penalties, all of which would be imposed by federal courts, may include jail sentences as well as fines. Again, these involve negligent or knowing violations.

Negligent violations are punishable by a fine of not less than $2,500 and not more than $25,000 per day of violation, and by imprisonment for not more than one year. Fines and prison sentences may be twice as high for repeat offenses.

Knowing violations carry penalties of $5,000 to $50,000 per day of violation and up to three years' imprisonment. Again, courts may double these penalties for repeat violations.

Knowing violations that place another person in imminent danger of death or serious bodily injury are punishable by fines of up to $250,000 per day and up to 15 years in prison. When the violator is a corporation, the fine may be increased to $1 million per day. Repeat offenses are subject to doubled fines and imprisonment.

The act creates a separate class of violations for parties that file false statements or tamper with monitoring equipment. Violations are punishable by fines of up to $10,000 and jail sentences up to two years, with penalties doubled for repeat offences.

In theory, a company could be subject to enormous fines if long-term violations are discovered. For example, in November 1990 Weyerhaeuser pleaded guilty to criminal violations of the Clean Water Act after it was discovered that the company had been discharging contaminated water into a tributary of Washington's Chehalis River—on a daily basis for nine years. Under the plea agreement, however, Weyerhaeuser pleaded guilty to just five counts of misdemeanor (negligent) violations of the Clean Water Act and was fined the maximum $25,000 on each count for a total of $125,000. The company also agreed to pay $375,000 into a trust fund dedicated to improvement of water quality in the area—an increasingly common remedy (*U.S. v. Weyerhaeuser Co.*, W.D. Wash., No. CR. 90-298S, Nov. 16, 1990).

Once the violations were discovered, Weyerhaeuser took immediate action to correct the problem, which only involved about 20 gallons of wastewater per day. This probably convinced the court to accept the company's five-count plea.

Courts are tougher on companies that knowingly violate the law, as was the case in *U.S. v. Menominee Paper Co.* in June 1990. The company and prosecutors agreed on a $2.1 million penalty in a case that involved discharge violations and inaccurate monitoring and reporting.

In another recent case of knowing violations of the Clean Water Act, Eagle-Picher Industries Inc. agreed to pay $1.5 million under a consent decree for discharging cadmium, chromium, mercury, zinc, and other pollutants into Missouri's Lone Elm Creek without a permit. The company also was charged with violating pretreatment requirements in its discharges to the Joplin, Missouri, municipal sewer system (*U.S. v. Eagle-Picher Industries Inc.*, (W.D. Mo., 30 ER Cases 1798 (1990)). The first conviction under the Clean Water Act's knowing endangerment provisions came in November 1990 in *U.S. v. Borjohn Optical Technology Inc.*, No. Cr. 89-256-WD (D. Mass., Nov. 7, 1990)). John Borowski and his company were convicted of ordering employees to dump untreated nickel plating wastes into the Burlington, Massachusetts, public sewer system. Prosecutors charged that the nitric acid in the wastewater could cause serious burns to respiratory tracts and other exposed areas.

Borowski was sentenced to 26 months in prison, placed on two years' probation, and ordered to pay a $400,000 fine. His company was ordered to pay another $50,000 fine and also to pay a lump-sum medical insurance premium to cover two employees who were exposed to the wastes.

The highest fine ever levied against a city for pretreatment violations came on July 2, 1990, in a suit against El Paso, Texas. The city's pretreatment program was approved by EPA in 1982, yet federal prosecutors charged that El Paso failed to identify industrial users, issue permits, inspect and monitor factories, or establish local limits (*U.S. v. El Paso*, No. E-P89 CA347 (W.D. Texas (7/2/90)).

Civil suits can be filed by citizens groups as well as the federal government. In July 1990, the Public Interest Research Group of New Jersey reached a consent agreement with Exxon U.S.A. that called for Exxon to spend $2.8 million to reduce the amount of petroleum-contaminated wastewater generated at its plant in Bayonne, New Jersey. Exxon also agreed to pay the group's legal expenses (*Public Interest Research Group of New Jersey Inc. v. Exxon U.S.A.*, (D.C. N.J., No. 89-2284).

Citizens' suits came into wide use during the 1980s, as environmental and public interest groups learned the ropes of using the discharge monitoring reports (DMRs) submitted by NPDES permit holders to determine whether treatment standards were violated. Although searching DMRs for violations provided a relatively easy first step in identifying problems, the citizen plaintiffs found they had to clear other hurdles to successfully prosecute an enforcement action:

Before filing in a federal district court, citizen plaintiffs first must provide 60 days' notice to the alleged violator, to EPA, and to state authorities about the nature of the threatened suit.

If the federal or state government then decides to proceed with its own enforcement action, this "diligent prosecution" will preclude the citizen from going forward with the suit.

Should the citizen go ahead and sue, the next barrier is to show that the citizen had legal standing to pursue the action. This means that the citizen plaintiff has to show that the alleged violations affected a waterway that was used by the citizen or by members of the group filing the action.

The plaintiff also has to show a "good faith" belief that the violations were ongoing—that they were continuing at the time the suit was filed.

After these criteria have been met, the suit may proceed, with the citizen plaintiff trying to prove violations based on the evidence contained in the DMRs. If the federal court decides that the facility was indeed liable for discharge permit violations, the court may fine the violator, with penalties going to the U.S. Treasury, and issue an injunction against further violations.

Usually, these citizen actions end in settlements, with a growing trend toward having the alleged violator pay for programs to benefit the damaged waterway through studies, public education, or cleanup projects.

III. SAFE DRINKING WATER ACT

INTRODUCTION

The water pollution control permit system discussed above is designed to protect surface waters. Most drinking water, however, comes from groundwater sources. These are generally known as aquifers: porous underground rock formations that hold water. They are susceptible to contamination, just as surface waters, and are protected by the Safe

Drinking Water Act, last amended in 1988 (Public Law 100-572).

WELLHEAD PROTECTION

A key to protection of groundwater was the establishment, in the 1986 amendments, of a wellhead protection program. Under this program, states are given grants by EPA to designate "wellhead protection areas," defined as "surface and subsurface area[s] surrounding a water well or wellfield supplying a public water system, through which contaminants are reasonably likely to move toward and reach such water well or wellfield" (Section 1428(e)).

As is the case with most environmental programs, it is up to states to develop these wellhead protection programs, subject to EPA approval and certain minimum standards outlined in the law. States must designate wellhead protection areas "based on all reasonably available hydrogeologic information." They must identify all potential manmade sources of contamination in those areas and adopt control measures to protect the water supply from contamination.

State wellhead protection programs must specify the technical and financial assistance, education, training, and demonstration activities needed to protect water supplies. States also are required to develop contingency plans and identify alternate sources of drinking water in the event of contamination.

SOLE-SOURCE AQUIFERS

State and local governments may apply, through the governor, for grants to protect any "sole-source aquifer," defined as "the sole or principal source of drinking water for the area and which, if contaminated, would create a significant hazard to public health . . . ." (Section 1424(e)). The EPA administrator may, on his or her own initiative or on petition by a state or local government, designate such areas, and that decision is announced in the *Federal Register.*

Once a sole-source aquifer is designated by EPA, no federal assistance of any kind may be used to support a project that has potential to contaminate that aquifer. Often the protection covers the aquifer's "recharge zone," the area through which water flows from the surface into groundwater.

STANDARDS

"Safe drinking water is not simply a social or environmental good, it is a necessity to life." So wrote David W. Schnare of EPA's Office of Drinking Water in a 1984 policy review. In EPA's 1987 *Unfinished Business* report and a December 1989 follow-up, *Comparing Risks and Setting Environmental Priorities*, the agency declared that contamination of drinking water represents one of the most serious potential threats to human health, far outweighing the hazards of toxic waste sites or local air pollution.

With this philosophy in mind, EPA has based its drinking water standards on human health effects (as opposed to a standard based on BAT), attempting to limit contamination to a "no observed adverse effects level." These numeric standards are known as "maximum contaminant levels," or MCLs.

Drinking water standards are important for another reason: they form the basis for hazardous waste cleanup standards under the Resource Conservation and Recovery Act and the Comprehensive Environmental Response, Compensation, and Liability Act (CERCLA). Those statutes complement the Safe Drinking Water Act by limiting or preventing hazardous waste disposal practices that could pollute groundwater.

When responsible parties are forced to clean up contaminated groundwater at sites covered by the Resource Conservation and Recovery Act or CERCLA, the remedy often becomes very expensive. This is particularly true if the parties must pump out the contaminated water, treat it, and return it to the aquifer.

The agency's drinking water standards may take on additional influence if states use them to regulate pesticides or toxic chemicals, as some chemical industry officials predict.

RESPONSIBILITIES OF DRINKING WATER SUPPLIERS

Whether drinking water comes from groundwater or surface water, suppliers must meet certain standards. They must control levels of the inorganic chemicals arsenic, barium, cadmium, chromium, lead, mercury, nitrate, selenium, and silver.

Suppliers must prevent contamination from three classes of organic compounds as well: chlorinated hydrocarbons, chlorophenoxy compounds, and total trihalomethanes. They also must limit human exposure from radionuclides—any radioactive particles—to 4 millirem per year and keep radium-226 and radium-228 levels to 5 picoCuries per liter.

Many chemicals are removed through an *ion-exchange* process. The goal is to convert them from soluble chemicals to forms that do not dissolve in water. A substance such as alumina often is used as a coating on a filter, and as the water passes through the filter the contaminant forms a salt (or other compound) that remains on the filter.

A less-often-used treatment process is *reverse osmosis*. This involves a membrane (sometimes combined with an electrical charge) that promotes the migration of contaminants to one side of a chamber. While this process is effective, it also is slow and expensive.

Some volatile chemicals can be removed by pumping hot air or steam into the contaminated water. This is known as *air stripping* or *steam stripping*. Scientists are hard at work developing new treatment processes, including those that remove radionuclides from water.

The ideal way to provide clean water, of course, is to prevent contaminants from entering the system in the first place. This is where the wellhead protection program comes in.

Lead contamination poses the greatest challenge for drinking water authorities because it has been used so widely in plumbing systems, mostly in lead solder that connects copper and brass pipes. It not is illegal to use lead solder in plumbing.

Much of the lead contamination in drinking water comes not from the suppliers but from individual domestic and commercial plumbing systems. For this reason, drinking water authorities have strenuously opposed any lead standard based on measurement at the consumer's tap, arguing that they have no control over those lead levels. A compromise has been reached whereby EPA will require suppliers to monitor consumer taps and institute water corrosivity controls if lead levels exceed 15 parts per billion at any point in the system.

Corrosion control could involve a reduction in the acidity of the water or addition of a substance such as orthophosphate that would form a protective coating inside pipes.

Another major responsibility of drinking water authorities is to prevent microbiological contamination, which generally is caused by improper or inadequte sewage treatment and discharge. It also can be caused by animal wastes. In addition, some naturally occurring water-borne disease can cause human health problems.

Microbiological contamination is controlled by monitoring levels of coliform bacteria, the class of organisms whose presence points to inadequate sewage treatment. All suppliers must monitor for coliform bacteria at least once a month—the requirement involves a sliding scale, and a system serving 4 million people must collect at least 480 samples per month! Coliform and other bacteria are controlled through disinfection, usually by chlorine.

Finally, suppliers must keep water clear of silt and other solids, thus, EPA has developed standards based on turbidity units. A drinking water authority must meet a standard of five turbidity units unless it can demonstrate to the state that it does not interfere with disinfection. Turbidity is generally a problem only for drinking water from surface water sources. Filtration is typically used to eliminate turbidity.

Turbidity once was measured by holding a candle behind a container of water. Today a more objective scientific test is used, although the principle—whethr light can penetrate the water—remains the same.

UNDERGROUND INJECTION

Underground injection wells are designed to inject wastewater so deeply that it cannot contaminate any aquifers—typically far below the lowest aquifer. They are used primarily to dispose of aqueous hazardous waste and are regulated under Part C of the Safe Drinking Water Act.

Any underground injection of waste requires a permit. However, EPA is directed by the act not to interfere with oil and gas drilling. The agency cannot prohibit reinjection of brine or other fluids brought to the surface during exploration, nor can it prohibit injection of fluids intended to recover oil from a well, unless it can demonstrate a clear threat of drinking water contamination.

One of the most common problems associated with underground injection is a leak in the walls of the well itself. Ideally, the well should prevent any migration of fluids into the soil until those fluids reach the bottom of the well, which would be beneath an impermeable layer of soil.

ENFORCEMENT

Enforcement of the Safe Drinking Water Act essentially parallels the Clean Water Act. If a state fails to take appropriate action against a violation within 30 days, EPA is empowered to issue an administrative order or file a civil suit on its own.

Violations are subject to civil penalties of up to $25,000 per day. Fines totaling less than $5,000 may be imposed by the agency administratively, without any court action. Fines above $5,000 require federal district court approval. Jail terms of up to three years are authorized for willful violations of the act.

The criminal sanctions were added to the Safe Drinking Water Act by Congress in 1986. The first criminal prosecution came in November 1990 when a U.S. Grand Jury indicted Weatherbee Supply Inc. of Bowling Green, Kentucky, and its president, Glenn Weatherbee. Weatherbee was charged with 10 counts of willfully violating the act by constructing and operating five underground injection wells without permits. He also was charged with filing false information in a permit application.

Weatherbee faced a maximum prison sentence of more than 30 years and a penalty of up to $2.75 million. The company could have been fined twice that, up to $5.5 million.

Charges against Glenn Weatherbee eventually were dropped, however, in exchange for a guilty plea by the company. Weatherbee Supply was fined $7,500 and put on five years' probation (*U.S. v. Weatherbee*, No. CR-90-00019-BG(M) (W.D. Kan. Apr. 15, 1992)).

Chapter 4

TRANSPORT & IDENTIFICATION

I. THE HAZARDOUS MATERIALS TRANSPORTATION ACT

INTRODUCTION

Transportation of chemicals and hazardous materials is regulated under the Hazardous Materials Transportation Act (HMTA). The law, enacted in 1975 and revised in 1990, gives the Department of Transportation (DOT) authority to regulate materials that may pose an unreasonable risk to health, safety, property, or the environment during transport by air, water, highway, and rail.

Prior to passage of HMTA, rules were written as the U.S. transportation system evolved. Rail transportation was regulated first, providing a model for restrictions on other modes of transport. Explosives, flammable liquids, and other hazardous materials were among the first substances regulated.

When HMTA was enacted in 1975, it consolidated decades of hazardous materials rules issued for each mode of transport. Under HMTA, DOT can:

- Define a substance or a type of material as hazardous when transported in commerce (Section 104)
- Regulate any safety aspect of hazardous materials transportation, such as packing, repacking, handling, labeling, marking, placarding, and routing (Section 105)
- Regulate all aspects of package manufacturing, maintenance, and testing (Section 105)
- Issue exemptions from any rules, if an alternative is available that would equal or exceed the level of safety in the rule or if the exemption is in the public interest (Section 107)
- Enforce the law through inspections, subpoenas, or hearings (Section 109), levy civil penalties (Section 110), seek legal injunctions, and declare imminent hazards (Section 111)
- Preempt state laws that are inconsistent with HMTA (Section 112)

Shippers, brokers, carriers, transporters, and those who make, recondition, repair, or test containers for hazardous materials transport are covered by HMTA. They are required to register with DOT and pay a registration fee. Responsibility for complying with HMTA is spread among the shipper of the material, the carrier of the shipment, and the manufacturer of the shipping container or packaging material.

The law directs DOT to classify materials by hazard. Shippers and carriers must comply with DOT regulations ranging from how to prepare a shipment for transportation to what modes of transportation are acceptable for a specific type of hazardous material. These requirements are summarized in the Hazardous Materials Table, which lists more than 17,000 substances and sets standards for shipping papers, package marking, labeling, and vehicle placarding. Hazardous substances include explosives, radioactive materials, etiologic agents (disease-causing to humans), flammable, combustibles, poisons, corrosives, organic peroxides, oxidizers, compressed gases, and irritants. Hazardous wastes and other substances that are regulated under other environmental laws, such as the Resource Conservation and Recovery Act (RCRA) and the Comprehensive Environmental Response, Compensation, and Liability Act (CERCLA), are regulated by DOT when they are transported.

All types of carriers—private, common, and contract haulers—must comply with DOT's general carrier rules and mode-specific rules. Carrier regulations govern loading, unloading, special labeling, inspection procedures for drivers, and hazard class restrictions. Carriers must report accidents and releases of hazardous materials to the appropriate federal or state agencies. For example, many types of releases must be reported to the National Response Center.

DOT requires employees in safety-sensitive positions in all modes of transportation, such as railroad crews, aircraft and vessel operating and maintenance staff, and drivers, to be tested for drug use. Testing can be required for employment, in postaccident evaluations, during periodic checks, under random programs, and when reasonable cause exists.

Exports of hazardous materials and hazardous waste leaving the United States are subject to regulations of other countries and those of international regulatory agencies. U.S. hazardous materials packaging rules are based on the international packaging standards of the United Nations Committee of Experts on the Transport of Dangerous Goods and the Technical Instructions of the International Civil Aviation Organization (ICAO).

DOT revised its hazardous material transport rules in December 1990 to make the requirements consistent with the United Nations recommendations. Implementation of DOT's so-called HM-181 standards (from "Hazardous Materials Docket No. 181") will be phased in throughout the 1990s.

In the 1990 reauthorization of HMTA, Congress delineated the areas over which the federal government has exclusive jurisdiction. States can gain authority to enforce hazardous materials transport standards, however, by adopting rules that are "substantially the same" as the federal regulations. State rules that are not substantially the same are preempted by the federal standards.

HMTA empowers the federal government to assess civil and criminal penalties for violations. DOT also can issue imminent hazard orders to suspend or restrict the transport of a material that may cause serious injury or death. The maximum civil fine is $25,000 for each violation and each day of continuing violation. The minimum fine is $250. The maximum criminal fine is $250,000 for an individual and $500,000 for a firm and/or five years' imprisonment.

Enforcement and compliance for hazardous materials shipped by a single mode, such as highway only, are governed by the individual administrations—Federal Aviation Administation (air), Federal Highway Administration (road), Federal Railroad Administration (rail), and U.S. Coast Guard (water). DOT's Research and Special Programs Adminstration (RSPA) enforces rules applicable to multimodal transport.

SHIPPERS

Hazardous materials shipments must be packaged, labeled, and transported according to the hazards associated with the materials. Shippers, including brokers and freight forwarders, play a pivotal role in ensuring compliance. HMTA defines shippers as those who offer hazardous materials for transportation or anyone who performs the shipper's packaging, labeling, or marking functions. Under HMTA, shippers must:

- Classify and describe the shipment according to the hazard type (49 CFR 173)
- Mark the shipment according to the hazard type (49 CFR 172)
- Determine that the packaging or container was made, assembled, and marked according to DOT rules or exemptions (49 CFR 173)
- Bring packages into compliance with DOT packaging specifications and tank-car specifications (49 CFR 178, 179)

- Notify appropriate parties, such as state officials, of the dates of shipments of certain materials, including radioactive materials, by approved highway routes (49 CFR 173)

Compliance with HM-181, which revised DOT rules for hazard communication, classification, packaging, and handling, will be phased in during the 1990s. Until the rules take effect, shippers may comply with either the old rules, the revised rules, or a combination of old and new rules.

Hazard Classification

Before hazardous materials can be shipped, the shipper must assign a hazard classification to the substance. The classification informs everyone who handles the shipment what to do in an emergency. The classifications are codified at 49 CFR 173, Subpart A.

There are eight hazard classes, two catch-all classes, and 15 subclasses:

- Class 1—Explosives
- Class 2—Flammable gases
- Class 3—Flammable liquids (any liquid that has a flash point of less than 60°C (140°F))
- Class 4—Flammable solids (materials that are explosive, combustible, or dangerous when wet)
- Class 5—Oxidizers (materials that produce oxygen that can cause or enhance the combustion of other material)
- Class 6—Poisonous materials (materials, other than gas, that are known to be so toxic to humans as to be a hazard to health during transportation, including etiologic (disease-causing) agents)
- Class 7—Radioactive materials
- Class 8—Corrosive material (liquids or solids that cause visible and irreversible alterations in human skin tissue at the site of contact, or liquids that have a severe corrosion rate on steel or aluminum)

- Class 9—Miscellaneous hazardous materials (a material that presents a hazard during transportation but is not included in any hazard class)
- Other regulated materials (materials, such as consumer goods, that are subject to DOT regulations and pose a limited hazard during transportation due to form, quantity, or packing)

Although DOT allows shippers to comply with either the old or new rules during the transition period, shippers do not have the option of mixing the old and new requirements. The materials must be described in exactly the same manner on shipping papers as on packages, labels, and placards.

A material that meets the definition of more than one hazard class must be classified according to DOT's Precedence of Hazard Table, which sets up a hierarchy of classification characteristics (49 CFR 173.2a). Any shipper who cannot determine the classification for a substance from the Hazardous Materials Table can consult DOT rules at 49 CFR 173, Subparts C through I.

Once a shipper selects the proper classification for the materials, the DOT Hazardous Materials Table is used to determine the most accurate shipping name for the substance. One or more of DOT's three packing groups are assigned to materials corresponding to their hazard class and proper shipping name. The three groups indicate the relative degree of hazard of the material being packaged for transport: Packing Group I, great danger; Packing Group II, medium danger; and Packing Group III, minor danger.

Hazardous Materials Table

The primary guide to hazardous material classification is DOT's Hazardous Materials Table (49 CFR 172.101). The table also is used to determine the packaging requirements for the materials. DOT revised the table in 1990 as part of the HM-181 process.

The table lists more than 17,000 substances. It consists of 10 columns ranging from proper shipping name to hazard classification. The table also directs the user to DOT requirements on labeling, packaging, quantities aboard aircraft, and stowage of hazardous materials on vessels.

The table contains the following information on all listed materials:

- Symbols designating special requirements such as unique restrictions for transport of materials by air, water, only within the United States, or international transportation
- Proper and preferred shipping name
- Hazard class
- Identification number
- Packing groups that are appropriate for the material and are associated with the degree of danger presented by the material
- Labeling requirements to warn of hazards associated with the hazard classification and proper shipping name
- Special codes for unique packaging requirements associated with the material during single mode transport
- Special numeric designations for multimodal shipments
- Packaging requirements, nonbulk packaging, bulk packaging, and exceptions
- Quantity limitations per package for rail and air transport, or prohibition against transport by rail or air
- Vessel stowage locations on board cargo and passenger boats, and other provisions for water transport

An appendix to the table contains the Hazardous Substance and Reportable Quantities (RQ) list. This alphabetical list of substances is designed to be used in conjunction with the Hazardous Materials Table if a spill occurs during transport. Releases exceeding the RQ threshold must be reported to the National Response Center.

Packaging, Containers

HM-181 shifted packaging requirements from specification-based rules to a more flexible, performance-based set of restrictions. The shipper is responsible for determining whether a container complies with performance and testing standards when preparing hazardous materials for transport by air, highway, rail, or water. Manufacture of "specification" packaging will be prohibited after October 1, 1994. Use of such packaging will be prohibited after October 1, 1996 (49 CFR 173, Subpart B).

Performance-Oriented Packaging

HM-181 eliminated about 100 DOT specifications for nonbulk package construction and introduced 20 United Nations performance-oriented standards (49 CFR 178). The U.N. standards, as embodied in HM-181, require that containers meet tests for structural integrity. The revised packaging rules do not assign specific types of packages for specific hazardous materials. The performance standards are based on general construction guidelines and list tests each package must meet based on the risk level of the material to be shipped. The risk level is reflected in the testing requirements for each of the three packing groups.

Under the revised packaging scheme, package manufacturers may need to subject their products to one or more of the following performance tests: the drop test, the leakproofness test, the stacking test, or the cooperage test for bung-type wooden barrels. For example, a steel drum must pass two drop tests if it is designed to contain materials that pose a "great danger" and for which its maker wants approval as a Packing Group I container. In these tests, drums containing solids or liquids are dropped from a height of 1.8 meters and are oriented to fall and strike certain points on the drum. The tested drums are then evaluated.

Container manufacturers must clearly mark their products to certify that the container complies with DOT regulations. The containers must be marked in a way that is accessible, easily read, and permanent.

One of three DOT packing groups is assigned to each hazardous material listed in the Hazardous Materials Table. Special packing provisions apply to Class I explosives and Class 7 radioactive material, and they are not assigned a packing group. The packing group number indicates the level of danger presented by the chemical.

General Packaging Requirements

DOT has three basic criteria for hazardous material packing (49 CFR 173.24). Shipping containers should be designed, constructed, maintained, filled, and closed, so that under normal transportation conditions (1) the contents will not be released into the environment in identifiable amounts, (2) the package will not be damaged significantly, and (3) heat or pressure will not build up to the point of explosion through vapor or gas mixtures.

These requirements apply to bulk, nonbulk, new, used, specification, and nonspecification packaging. Packing requirements apply to all modes of transport unless specifically stated or exemptions are authorized. All packages of hazardous materials must be open to inspection by authorized representatives of the initial carrier or a DOT representative.

Exceptions

DOT can grant exemptions from packaging rules for small quantities of hazardous materials and for some hazardous wastes.

Some Class 3 flammable liquids, Class 8 corrosive materials, and Class 9 miscellaneous materials are eligible for the small-quantity exception. Small-quantity packaging still must pass certain tests, however (49 CFR 173.4). Certain hazardous wastes are exempt from DOT packaging requirements, but such exemptions may apply only to wastes transported by highway (40 CFR 173.12).

Forbidden Materials, Packaging

DOT has listed 10 types of materials or situations in which packaging and transport is forbidden (49 CFR 173.21). Shippers cannot transport:

- Chemicals that are designated "forbidden" in column 3 of the Hazardous Materials Table
- Explosives that have not been approved for transport by DOT
- Electrical devices that are likely to create sparks or a dangerous amount of heat
- Items transported by air whose magnetic fields exceed DOT limits
- Combinations of unstable hazardous materials, including those that produce excessive heat, release dangerous gases or vapors, and degrade or polymerize above certain temperatures
- Packages that will detonate in a fire (other than explosives)
- Packages containing a cigarette lighter or other device combining an ignition element and fuel, unless inspected by the Bureau of Explosives and approved by DOT
- Ketone peroxides that contain less than 9 percent oxygen

Container Reuse

Hazardous material containers may be reused if they meet DOT requirements. Before each reuse, DOT requires that the packaging be inspected. A package may not be reused if any incompatible residue, rupture, or other damage is present that will reduce the structural integrity of the packaging, unless it is reconditioned according to DOT rules (49 CFR 173.28).

Metal and plastic containers must meet minimum thickness requirements to qualify for reuse. DOT also has reconditioning standards for metal drums and other non-bulk packaging. Shippers may not reuse packaging made of paper, plastic, film, or textile. Reuse of nonbulk packaging for the transport of hazardous waste is also prohibited.

Some nonbulk packaging may qualify for remanufacture, the process of converting packaging that meets one standard to one that meets another standard. The process also includes the replacement of integral structural parts. Remanufactured nonbulk packaging must meet the requirements of 49 CFR 173, Subpart L.

Miscellaneous Provisions

DOT has established separate packaging requirements for transport of hazardous materials in special circumstances.

Leaking packages (49 CFR 173.3) may be placed in metal or plastic salvage drums if certain conditions are met.

Agricultural chemical shipments (49 CFR 173.5) may not have to meet the hazard class packaging requirements if they are transported in quantities of less than one case lot or are repackaged, although other conditions must be met.

Department of Defense (DOD) hazardous materials (49 CFR 173.7), including commercial shipments associated with a DOD contract, must be packaged in containers that are of equal or greater strength or be certified according to military specifications.

Shippers of cryogenic materials (refrigerants) must have a registration statement for transporting low-temperature, flammable materials (49 CFR 173.11).

Identification, Labels, Placards

Shippers that intend to transport hazardous materials in vehicles, freight containers, or packages must ensure that each is properly identified. On a properly marked package, the identification number of the material is clearly visible. The marking requirements vary depending on the type of material (e.g., liquid) and method of transport (e.g., bulk, nonbulk) (49 CFR 172, Subpart D).

In general, marking must be durable, written in English, and printed on or affixed to the surface of a package or on a label, tag, or sign with contrasting color. It must be legible and set apart from other markings (such as advertising), and it should not appear on the bottom of the package.

Nonbulk packaging must be marked with the proper shipping name and DOT identification number, preceded by "UN" (United Nations) or "NA" (North America) as appropriate. DOT defines nonbulk packages as those with a capacity of 450 liters (119 gallons), 400 kilograms (882 pounds), or a water capacity of 435.6 kilograms (1,000 pounds). Packages must be marked with the technical name in parentheses along with the proper shipping name (from the Hazardous Materials Table). If the package is authorized by an exemption, it must be marked "DOT-E" followed by the exemption number.

Bulk packaging also must be marked with the appropriate shipping name and identification number. Because bulk shipments are large, DOT created separate marking requirements. A train tank car and a cargo tank truck are examples of bulk shipping methods. In these cases, the identification number must be marked on each side or opposing sides of the packaging or on all sides of a motor vehicle carrying permanently installed cylinders. The outside of a bulk package subject to an exemption must be plainly and durably marked "DOT-E" followed by the assigned exemption number. Bulk packaging usually must remain marked when it is emptied.

Identification Numbers

Identification numbers must be displayed on a transport vehicle or freight container containing hazardous materials that does not require hazard-warning placards. DOT's emergency response handbook lists the numbers, the identity of the material, and emergency response information. The numbers are listed in the Hazardous Materials Table.

The numbers, which are used internationally, must be displayed either on orange panels or on a white square-on-point display configuration that has the same dimensions as a placard. DOT has special provisions for materials in hazard classes for which placards are specified and identification number displays are required. If an identification number is lost, damaged, or destroyed during transportation, the carrier must replace all the missing or damaged numbers as soon as possible.

Labels

HMTA requires shippers to label packages or containment devices with labels specified for the material in the Hazardous Materials Table. Labels are hazard warnings, usually symbols. DOT has a label for each hazard class and subclass (49 CFR 172, Subpart E).

Labeling requirements depend on the contents of each package. If, for example, a package contains materials that fall under more than one hazard class, it must bear labels representing each class. Column 6 of the Hazardous Materials Table contains labeling information. A label that meets U.N. specifications may be used in place of the corresponding DOT label.

DOT specifications for labels address their durability, design, size, and color. The label cannot be obscured and must be attached to a surface other than the bottom of the package. When multiple labels are required, they must be placed near each other. Some large, nonbulk packages and some portable tanks and freight containers require duplicate labels because of their size. In these cases, labels must be displayed on at least two sides of the package.

Exemptions to labeling rules are authorized when:

- Certain nonpoisonous gases are durably and legibly marked
- Military explosives are shipped by or on behalf of the DOD

- Packages of DOD hazardous materials, other than ammunition, are loaded, unloaded, and escorted by DOD personnel
- Compressed gas cylinders are permanently mounted on a transport vehicle
- Freight containers, aircraft, or portable tanks are appropriately placarded
- Radioactive materials are transported in accordance with other DOT rules

Exemptions for shipments of small quantities of hazardous materials also may be obtained. (See Chapter 3 for a discussion of small-quantity exemptions.)

Placards

Each bulk packaging, freight container, motor vehicle, rail car, or other loading device used to transport hazardous materials must bear a placard that identifies the contents and hazard. A placard is a color-coded, diamond-shaped sign with symbols, numbers, and other information to communicate the hazard (49 CFR 172, Subpart F).

Shippers and carriers share responsibility for ensuring that a shipment is properly placarded before transport. Shippers must provide the required placards to the carrier handling the shipment, and the carrier is prohibited from accepting shipments that are not properly placarded. When more than one placard is required for certain nonbulk packaging, the freight container, vehicle, or rail car can be placarded with the message "DANGEROUS," instead of separate placards for each material.

Placards placed on motor vehicles or rail cars must be readily visible. Placards may be placed on the front of a truck in addition to or instead of placarding the front of the cargo body. Each placard must be attached securely or placed in a DOT-specified holder; located away from ladders, pipes, or other obstructions; placed where dirt or water from the vehicle cannot reach it; isolated from other markings; displayed so that words and identification numbers are read from left to right; protected from damage; and affixed to a background of contrasting color.

Exceptions to the DOT placarding requirement include shipments of infectious substances, material classed as "other regulated material," materials packaged under the requirements for small quantities, and combustible liquids in nonbulk packaging. In addition, shipments that conform to the recommendations of the U.N. Committee of Experts on the Transport of Dangerous Goods may be used in place of DOT regulations.

Shipping Papers

HMTA requires most shipments of hazarous materials to be accompanied by shipping papers or a hazardous waste manifest. Shipping papers contain a summary of the material's hazards and the associated packaging, marking, and other information. Shippers must prepare the papers and give them to the carrier. In most cases, shipping papers are combined with other documentation that accompanies any package shipped by a carrier (49 CFR 172, Subpart C).

DOT requires shipping papers to include information about the type of packaging, proper shipping name, hazard class, and emergency response information. Depending on the mode of transport, there may be specific requirements for shipping papers. The Federal Highway Administration, for example, has established safety-related requirements for all shipments.

DOT specified exactly how the information should be arranged on the shipping papers. The required format follows the order of the 10 columns in the Hazardous Materials Table. In addition to the basic shipping information, each mode of transportation has specific requirements for descriptions of hazardous materials. The following are examples of shipping descriptions:

- Gasoline, 3, UN 1203, PG II

- Motor fuel antiknock compound (tetraethyl lead), 6.1, UN 1649, PG I
- RQ, Allyl alcohol, 3, UN 3077, PG I
- Corrosive liquid, n.o.s. (caprylyl chloride), 8, UN 1760, PG II
- Flammable liquids, n.o.s. (contains xylene and benzene), 3, UN 1993, PG II
- Residue last contained naptha, 3, UN 1255

The proper shipping name (Column 2) is followed by the hazard class or division number (Column 3). If the proper shipping name is generic, the abbreviation n.o.s. ("not otherwise specified") is included on the shipping paper.

Next should be either the United Nations (UN) or North American (NA) identification number (Column 4), the packing group (Column 5) preceded by the letters "PG," and the total quantity of hazardous material covered by the description. An emergency response telephone number also must be listed on the shipping paper. The number need only be operational when the hazardous materials are in transit. The telephone number may be the number of an agency or organization capable of and responsible for providing information about the materials. Members of the Chemical Manufacturing Association can use CHEMTREC ((800) 424-9300) for emergency assistance.

All descriptions must be written in English, and abbreviations are not allowed unless specifically authorized by DOT. Multiple shipping papers are accepted as long as each page is numbered consecutively and the total number of pages is indicated.

If the shipment contains a limited quantity of a hazardous material, labeling and packaging exceptions apply. A very small quantity of a substance, such as a half pint of benzene, may qualify for the limited-quantity exception. Each shipping paper issued in connection with a shipment made under an exemption must bear the notation

"DOT-E" followed by the exemption number. If the shipment contains poisonous materials, this information must be included in the shipping name or entry for hazard class.

DOT specified three ways to present information about shipments containing both hazardous materials and materials that are not regulated as hazardous under HMTA: (1) the description of the hazardous material can appear first on the shipping paper; (2) information about the hazardous material can be entered in a color that clearly contrasts with the shipping paper; and (3) the shipping paper can have a column labeled "HM," and those entries for hazardous materials could be marked off with the letter "X" or the letters "RQ" (reportable quantity), if appropriate.

Shipping Certification

Shippers must certify that the hazardous materials being transported comply with DOT regulations, unless they are exempt (49 CFR 204). If the shipper acts as a private carrier or the carrier provides a cargo tank, then no certification is required. A certification statement must appear on the shipping papers. DOT has suggested two statements that are acceptable as a declaration of compliance:

- "This is to certify that the above-named [or herein-named] materials are properly classified, described, packaged, marked and labeled, and are in proper condition for transportation according to the applicable regulations of the Department of Transportation."
- "I hereby declare that the contents of this consignment are fully and accurately described above by proper shipping name and are classified, packed, marked and labeled, and are in all respects in proper condition for transport by [specify modes of transportation] according to applicable international and national governmental regulations."

The document must be signed by a principal, officer, partner, or employee/agent of the shipper. Shippers using air transport may substitute alternative language, specified for that mode of transportation. DOT does not require certification statements for an empty tank car that contains the residue from a hazardous material.

Hazardous Waste Manifest

All shipments of hazardous wastes must be accompanied by a Uniform Hazardous Waste Manifest (Environmental Protection Agency (EPA) Form 8700-22). The shipper is required to prepare the manifest and give it to the carrier. The manifest is required by RCRA (40 CFR 262, Subpart B). Copies of the manifest must be kept by anyone involved with the transportation of the waste. (See page 230 for an explanation of the RCRA manifest.)

If the manifest is completed according to EPA rules and DOT specifications (49 CFR 172.205), it may be used as the shipping paper.

The manifest must be signed by the waste generator, each carrier, and the receiving facility. Copies of the manifest must be kept by each party that signed the manifest for three years. The shipper (hazardous waste generator) also must keep the completed manifest for three years from the date the waste was accepted by the initial carrier. If the waste is being exported, the manifest must note the date of departure from the United States. DOT has established special provisions for handling the manifest if the hazardous waste is transported by rail.

Exemptions

Anyone who must meet requirements of HMTA may apply for an exemption. Exemptions are issued by the Hazardous Materials Safety Office of DOT. They are only issued if the applicant can prove that the proposed alternative would provide levels of safety that are consistent with the public interest and the Hazardous Materials Transportation Act. Sometimes DOT converts exemptions into permanent amendments or exceptions to its rules (49 CFR 107, Subpart B).

In 1989, DOT received 163 new exemption requests and 1,250 renewal requests. By the end of the year, there were 1,117 exemptions in effect. One exemption authorized the Olympic flame to be transported in the cabin compartment of a passenger airplane.

Although there is no form for exemption requests, DOT has prescribed what information must accompany a petition. Three copies of the exemption request must be sent to DOT in Washington, D.C. Unless the application is to remedy an emergency situation, applicants should allow at least four months for DOT to process the request. Applications must provide:

- The text of the regulation from which the exemption is sought
- The name, telephone number, and address of the applicant
- A detailed description of the proposal with all available supplemental information
- The chemical name, physical state, hazard classification, and other characteristics
- Relevant shipping and accident information
- Proposed mode of transport
- An explanation of why the alternative safety controls will meet DOT's requirements

DOT publishes a notice in the *Federal Register* announcing receipt of the application. Public comment may be received on the application. Another notice announces the approval or denial of the application. Applicants can withdraw their petitions prior to a final decision. In these cases, DOT publishes a notice announcing withdrawal of the application.

CARRIERS

All types of carriers—private, common, and contract haulers—must comply with DOT's general carrier rules and the mode-specific rules. Carrier regulations govern loading, unloading, special labeling, inspection procedures for drivers, and hazard class restrictions.

The carrier must ensure that hazardous materials are in proper condition for shipment. For example, the carrier must make sure that none of the packages are leaking or damaged. The carrier also is required to verify that the shipping papers, labeling, and marking are consistent.

Carriers must segregate certain materials. Substances that are incompatible and create a hazard if transported together are listed on segregation charts. Carriers also are responsible for seeing that their vehicle is properly placarded, that labels and other markings are not destroyed in transit, and that the cargo is properly loaded.

Congress confronted one issue of transporting incompatible materials by passing the Sanitary Food Safety Act of 1990. The act was designed to prevent truck drivers from carrying shipments of fruit juice and edible oils in the same tank car as formaldehyde or garbage without adequate cleaning between shipments.

DOT, in consultation with other federal agencies, is developing a list of nonfood products that can be safely transported and a second list of unacceptable food/nonfood combinations. The act also calls for the use of dedicated vehicles for the transport of materials (e.g., asbestos) that pose an extreme danger to human health. Penalties for violations of the act will be assessed in the same manner as those under HMTA.

Mode-Specific Carrier Rules

In addition to DOT's general carrier rules, all carriers must comply with mode-specific rules. These rules are designed to promote safety. There are four mode-specific agencies: the Federal Railroad Administration, the Federal Aviation Administration, the Federal Highway Administration, and the U.S. Coast Guard.

Rail Transport

Rail carriers may not accept shipments of hazardous materials unless they first receive a certified shipping paper to accompany the load and all other DOT rules have been observed. DOT has prescribed detailed requirements for handling explosives, gases, flammable liquids, flammable solids, and materials that are oxidants, poisonous, radioactive, or corrosive (49 CFR 174).

Within 48 hours of acceptance of hazardous materials for rail transport, the carrier must forward the shipment. Saturdays, Sundays, and federal holidays are excluded from the expedited shipment requirement. At transfer stations, rail yards, or other interchange points where only biweekly or weekly service is available, carriers must forward the hazardous materials on the first available train.

DOT has established a segregation table for materials that may not be loaded, transported, or stored together (49 CFR 174.81). The department also has specified the position in the train for a placarded rail car. Leaking packages may not be forwarded until repaired or reconditioned. Likewise, leaking tank cars may not be unnecessarily moved until the unsafe situation is corrected. DOT's general operating and handling rules impose other requirements:

- Each rail carrier must keep an extra supply of labels and placards to replace those that become lost or destroyed in transit.
- Each intermediate carrier or freight forwarder must keep a copy of the certified shipping paper for each shipment it handles.
- Each loaded placarded tank car must be inspected to ensure that it is in good operating condition before the carrier accepts it for transport.

- Notice must be provided to the train and engine crew of the location of rail cars loaded with hazardous materials.
- Train crews must have a copy of the shipping papers.
- Rail carriers must placard a tank car with an "overdue for testing" notice when required testing cannot be performed because the tank car is loaded.

Air Transport

Domestic air carriers must comply with HMTA restrictions for the transport of hazardous materials. DOT rules apply to materials aboard, attached to, or suspended from aircraft (49 CFR 175).

All air carriers in the United States that are registered for commercial air transport service must comply with DOT rules for loading, unloading, and handling hazardous materials. The department distinguishes between cargo-only and passenger-carrying aircraft when setting limits for materials aboard aircraft. Quantity limits for hazardous materials are listed in the Hazardous Materials Table.

The DOT air transport rules do not apply to government-owned-and-operated aircraft, private aircraft under "exclusive use" agreements with the government, and certain U.S. registered aircraft operated or leased by foreign nationals.

DOT exempted some materials that are otherwise subject to transportation rules, including hazardous materials used during flight for aerial seeding, fertilizing, or pest control; medical devices containing radiopharmaceuticals that are implanted in humans or animals; and carbon dioxide (dry ice) when no more than five pounds are used to refrigerate the contents of a package.

The DOT's general operating and handling rules also require air carriers to:

- Refuse hazardous materials for transport if they do not comply with DOT rules

- Report discrepancies to the nearest Federal Aviation Association Civil Aviation Security Field Office
- Notify the pilot-in-command of the hazardous material shipment and give the pilot the shipping papers before departure of the aircraft
- Maintain a supply of replacement labels to replace lost or damaged labels
- Inform passengers by prominently displaying a notice at ticket and baggage counters about hazardous materials restrictions.

DOT also specified procedures for loading, unloading, and handling hazardous materials on aircraft. DOT has created a stowage compatibility table to regulate the placement of packages that might react dangerously with one another in the event of leakage. In addition to cargo location specifications, DOT has established regulations for specific classifications of hazardous materials.

Water Transport

DOT and the U.S. Coast Guard share responsibility for regulating water transport. DOT regulates vessels, including barges and ferries, that transport hazardous material. The U.S. Coast Guard regulates other types of water transport and is responsible for enforcement of HMTA and other water transport safety laws (49 CFR 176).

Among the vessels covered by U.S. Coast Guard rules (and excluded from DOT's hazardous materials transport regulations) are tug and towing boats, vessels constructed only for the transport of flammable or combustible liquids, public vessels not used for commercial service, and boats used exclusively for pleasure.

Vessel carriers must comply with rules for shipping papers, including a requirement to prepare a dangerous cargo manifest. Carriers cannot accept hazardous materials for transport without obtaining a statement of certification from the shipper that the materials are properly classified, packaged, marked, and labeled. Carriers must maintain an adequate supply of labels to replace those damaged, destroyed, or detached.

The carrier, or duly designated agent, must inspect each hold or compartment containing hazardous materials at least once every 24 hours, weather permitting. The inspector must ensure that the cargo is in safe condition and no damage has occurred since loading and stowage.

DOT has established general segregation rules as well as specific requirements for cargo and passenger vessels. Segregation is achieved by maintaining certain distances between imcompatible hazardous materials or by requiring vessel construction (e.g., steel bulkhead or deck) between the materials.

DOT also has detailed loading, unloading, stowage, and segregation requirements for several classes of hazardous materials. If an accident involving hazardous materials occurs on board a vessel, the vessel master is responsible for ensuring the safety of the crew and/or passengers. In some emergency situations, the U.S. Coast Guard must be notified and a written report filed.

Highway Transport

Motor carriers must not accept a shipment of hazardous material unless they first receive a certified shipping paper and all other DOT rules have been observed. Carriers are responsible for properly placarding and labeling a vehicle intended to transport hazardous materials. DOT prohibits certain explosive materials and leaking or damaged packages to be transported by highway (49 CFR 177).

Prior to transport, carriers must ensure that tanks, barrels, drums, cylinders, or other packages that are not permanently attached to the vehicle are securely attached to the vehicle. If a shipment of a hazardous material is refused or cannot be delivered within 48 hours of arrival at the destination, the material must be disposed of by returning it to the shipper, placing it in storage, or selling or destroying the materials when necessary.

DOT has prescribed detailed requirements for loading, unloading, and handling certain classifications of hazardous materials. Carriers are responsible for informing the driver as to what routes to take when the vehicle contains Class 7 radioactive materials. Motor vehicles containing a "highway route controlled quantity" of radioactive materials must operate on preferred routes only. State agencies may select preferred routes to minimize radiological risk. For example, some states require trucks carrying radioactive materials to bypass densely populated areas.

DOT has specific requirements for the proper handling of accidents involving hazardous materials in transit. Motor carriers must comply with DOT's incident reporting requirements. (Incident reporting requirements are discussed below.)

Drivers transporting hazardous materials on public highways must comply with the general training and operating requirements as required by other statutes such as the Motor Carrier Act of 1980. That law, for example, requires carriers to maintain a minimum level of financial responsibility (49 CFR 387). The prime purpose of the financial responsibility provision is to ensure the public that the carrier can settle claims resulting from bodily injury, property damage, or environmetal restoration. The amount of liability coverage required depends on the kind and quantity of hazardous materials transported.

Spills, Emergency Response

Whenever hazardous materials are unintentionally released during transport, loading, unloading, or storage, HMTA mandates that the carrier report the incident to DOT. In addition to HMTA, some environmental statutes require releases of hazardous substances into navigable water, groundwater, earth, and air to be reported to the National Response Center ((800) 424-8802). The center is designed to provide hands-on assistance in a chemical emergency.

HMTA requires that carriers notify the response center if an accidental release results in injury requiring hospitalization, death, damages exceeding $50,000, evacuation, closure of major transportation arteries or facilities, or changing of a flight pattern. It also must be reported if it involves radioactive or disease-causing materials. Even if a release does not present a continuing danger to life at the scene of the incident, the release can be reported if, in the carrier's judgment, there is reason for concern. Each carrier making immediate telephone notification also must file a written report with DOT (49 CFR 171.15).

The U.S. Coast Guard and EPA are responsible for spill response and cleanup of oil and hazardous substances. DOT and EPA jointly enforce the reporting requirements. EPA requires immediate reporting of hazardous substances if the amount of accidental discharge exceeds the reportable quantity (RQ) as listed in the Hazardous Substances and Reportable Quantities Table. That table, which is sometimes called the RQ table, is an appendix to the Hazardous Materials Table (49 CFR 172.101) (see the earlier discussion on reportable quantities).

CERCLA mandates spill reporting and makes the carrier responsible for the cost of cleanup and for damages associated with a spill of hazardous materials or waste. If a truck carrying PCB-contaminated transformers overturns onto private property, the owner has the right to sue the carrier for damages. (See Chapter 6 for a discussion of CERCLA.)

Accidental releases of hazardous wastes are regulated differently. They must be cleaned up by the transporter and reported to the National Response Center or the appropriate agency as required by RCRA. The difference between RCRA and CERCLA is that RCRA regulates the handling of hazardous wastes (see the following discussion on RCRA's reporting requirements).

The Emergency Planning and Community Right-To-Know Act (EPCRA) also requires reporting of all releases of listed chemicals—both routine and accidental—above a certain threshold. Carriers can meet this notification requirement by telephoning the emergency number (911) or the operator (see the earlier discussion on EPCRA's reporting requirements).

Written Report

Although a carrier may have reported an incident by telephone immediately, DOT requires carriers to file a detailed written report within 30 days of the accidental discharge of hazardous materials or hazardous waste. If the report concerns a release of a hazardous waste, a copy of the Uniform Waste Manifest must be attached to the DOT report (49 CRF 171.16).

Two copies of the incident form (DOT Form F 5800.1) must be mailed to Information Systems Manager, DHM-63, Research and Special Programs Administration, DOT, Washington, DC 20590. Carriers must keep a copy of the report for two years.

Carriers who accidentally release certain low-risk hazardous materials are exempt from the written report requirement. These exemptions do not apply if the accident triggers DOT's requirements for immediate notification (i.e., the incident involves hazardous waste or occurs aboard an aircraft).

DOT Emergency Response Guidebook

To provide information for the immediate handling of spills, DOT and the Chemical Manufacturers Association have coordinated the development of the *Emergency Response Guidebook* and telephone information service called CHEMTREC.

The *Emergency Response Guidebook* (DOT Publication P 5800.5) is published by the Hazardous Materials Safety Office in DOT's Research and Special Programs Administration. The guidebook is intended to assist "first responders" (emergency response personnel) in making informed judgments during the initial phases of a transportation incident. Carriers also have the book in their vehicle. DOT intends to have a copy of the book in all emergency service vehicles in the United States.

CHEMTREC (Chemical Transportation Emergency Center) is a service of the Chemical Manufacturers Association, based in Washington, D.C. Shippers can hire CHEMTREC to provide emergency contact as required by DOT rules. The information service also may provide information to fire, police, and medical personnel about a chemical or know how to contact its manufacturer. CHEMTREC representatives can be reached at (800) 424-9300.

INTERNATIONAL RESTRICTIONS

Shipments of hazardous materials and hazardous waste leaving the United States are subject to regulations of other countries and international regulatory agencies. Barriers to international trade, such as conflicting packaging rules, was cited as one of the primary reasons for revising U.S. hazardous materials transportation rules in 1990.

Today, U.S. hazardous materials transportation rules are based on the internationally used packaging standards of the United Nations Committee of Experts on the Transport of Dangerous Goods and the Technical Instructions for the International Civil Aviation Organization. At least two other countries—Canada and Australia—have implemented domestic regulatory systems based on the United Nations recommendations.

European rules, which also are based on the U.N. recommendations, contained a provision that made non-U.N. packaging unacceptable for transport after December 31, 1991. As a result, use of pre-HM-181 packaging was unacceptable when exporting hazardous materials. To grant relief to exporters, DOT authorized compliance with the revised U.S. rules as of January 1, 1991. For transportation within the United States, the U.S. rules were scheduled to be phased in during a period of several years.

United Nations

Several documents concerning the transporation of hazardous materials have been developed by organizations affiliated with the United Nations:

- "U.N. Recommendations on the Transportation of Dangerous Goods" was written by the U.N. Committee of Experts, one of several subsidiary bodies to the Economic and Social Council. The council is indirectly supervised by the U.N. General Assembly.
- "Agreement Concerning the International Carriage of Dangerous Goods by Road," which is commonly called ADR, was written by the Inland Transport Committee. This committee is one of the U.N.'s Economic Commission for Europe committees.
- "International Maritime Dangerous Goods Code," an international code for water transport of hazardous materials, was written by the International Maritime Organization (IMO). IMO is a specialized agency of the United Nations that operates in an advisory capacity.
- "Regulations for the Safe Transport of Radioactive Materials" was issued in 1985 by a specialized agency of the United Nations, the International Atomic Energy Agency (IAEA Safety Series No. 6).
- "Technical Instruction for the Safe Transport of Dangerous Goods by Air" is the main product of the International Civil Aviation Organization, the U.N. group that governs air transport of hazardous materials.
- International agreements for rail transport of dangerous goods are administered by the Office Central des Transport Internationeau par Chemin de Fer.

STATE ROLE

HMTA authorizes states to regulate intermodal transport of hazardous materials so long as the rules are consistent with the federal rules. If a state rule interferes with the federal system, DOT will rule it inconsistent, and it will be preempted by the federal rules. Between 1977 and 1989, DOT issued 27 rulings on consistency issues (40 CFR 107, Subpart C).

Inconsistency rulings are based on an evaluation of whether the state rule would make it difficult for shippers and carriers to comply with both state and federal rules. This is called the obstacle test. DOT also evaluates whether it is possible to comply with both regulations; this is called the dual compliance test. The transportation department considers whether the state requirement reduces, increases, or leaves unchanged overall public safety.

Until 1990, DOT inconsistency rulings were advisory—binding rulings were made only by federal courts. When HMTA was reauthorized in 1990, the law was changed to make DOT rulings legally binding. A petitioner, however, can seek judicial review of an adverse DOT inconsistency ruling. Petitions for inconsistency rulings have been filed by states, highway carriers, rail carriers, and cities. For example, in 1982 the Nuclear Assurance Corp. asked for a ruling on New York State Thruway Authority regulations involving radioactive material transportation. DOT said the state's requirements were inconsistent with the federal rules.

Congress, in the 1990 reauthorization of HMTA, delineated several areas over which the federal government has exclusive jurisdiction:

- Hazardous materials classification, description, and designation
- Packing, repacking, handling, labeling, marking, and placarding of hazardous materials
- Shipping document preparation, execution, placement, and use
- Written reports about unintentional releases of hazardous materials
- Certified package design, manufacture, fabrication, marking, maintenance, reconditioning, repairing, and testing

The lawmakers also said states may adopt and enforce regulations addressing areas of exclusive federal jurisdiction only if they are "substantially the same" as the federal rules. If the state rules do not pass the substantially similar test, they are preempted.

One of the most contentious state-federal consistency debates occurred between the state of Ohio and four railroads. At issue was whether the Federal Railroad Safety Act preempted state hazardous materials rail transport regulations or whether the state regulations were consistent with HMAT.

The Supreme Court rejected a petition from the Ohio attorney general to make valid the state's regulations that a U.S. appeals court declared to be preempted. As a result, states cannot adopt and enforce their own rail hazardous materials rules. States can, however, participate in an enforcement program with the Federal Railroad Administration that allows state inspectors to check for compliance with federal rules (*Public Utilities Commission of Ohio v. CSX Transportation Inc.*, 111 S. Ct. 781, *cert. denied* January 22, 1991).

ENFORCEMENT

HMTA authorizes the federal government to assess civil and criminal penalties for violations of the act. The maximum civil penalty is $25,000 for each violation and each continuing day of violation. The minimum civil fine is $250. The maximum criminal fine is $250,000 for an individual and $500,000 for a firm and/or five years' imprisonment.

DOT also can issue imminent hazard orders to suspend or restrict the transport of a material that may cause serious injury or even death.

Most lawsuits under HMTA are handled by administrative proceedings, even though the government pursues certain civil cases and all criminal cases in the federal courts. There are three steps in administrative enforcement proceedings. First, a notice of probable violation is issued. Then, if the matter cannot be settled through negotiations, an order and possibly a penalty can be issued by the DOT chief enforcement counsel or a DOT administrative law judge. Most orders are issued by the chief enforcement counsel. The final step is to file an appeal of the order with the RSPA administrator.

Two types of civil enforcement cases are reviewed by U.S. district courts. The more common is a lawsuit filed by the government for failure to pay a penalty assessed in an administrative order. The second is an appeal of a decision by the RSPA administrator. There is nothing in HMTA that explicitly states this type of review procedure. Nonetheless, in the first 15 years of HMTA, there were at least three such federal court cases.

Enforcement and compliance monitoring of HMTA is shared among the federal modal agencies and RSPA. RSPA enforces the rules that are applicable to multimodal transport. The modal agencies within DOT are the Federal Railroad Administration (rail), the Federal Aviation Administation (air), the U.S. Coast Guard (water), and the Federal Highway Administration (highway). Lawsuits claiming a criminal violation of the law, however, are developed by DOT and agency officials and litigated by the Department of Justice. Most criminal cases are developed around alleged violations of general federal criminal laws, such as false claims (18 U.S.C. 287) and mail fraud (18 U.S.C. 1341). In the first 15 years after HMTA was enacted, only three HMTA-related criminal cases were prosecuted.

Although the penalties and types of enforcement actions are defined under HMTA, each modal agency has its own process, albeit similar, for initiating an administrative enforcement action:

- Rail transport enforcement actions include a notice of investigation that can lead to a final compliance order stating the required remedy for the violation (49 CFR 209).
- Most air transport violations are discovered by in-flight safety inspections, incident reports, or public complaints.

The Federal Aviation Association's process for initiating an enforcement action is codified at 14 CFR 13.

- Marine safety investigations in each of the U.S. Coast Guard's district offices are used to detect violations of HMTA.
- The Federal Highway Administration may initiate one of three types of proceedings to enforce HMTA: a driver qualification proceeding, a notice of investigation proceeding, or a civil forefeiture proceeding (49 CFR 386).
- Mutimodal enforcement by DOT includes issuing orders for immediate compliance, seeking a court-ordered injunction to prohibit a dangerous act, and assessing penalties based on the nature of the violation, degree of culpability, ability to pay, and history of prior offenses (49 CFR 107, Subpart D).

Examples of Administrative Proceedings

Below are examples of civil enforcement cases that were settled under the auspices of an adminstrative law judge. Fines were paid ranging from $175 to $97,000. A steamship company paid the $175 fine for allegedly completing a dangerous cargo manifest improperly, while a railroad paid a $97,000 fine because train crews allegedly did not have shipping papers.

Federal Railroad Administration

Union Pacific Railroad, a carrier, paid a $97,000 penalty for transporting hazardous materials without train crews having copies of shipping papers on more than one occasion, for improper descriptions on shipping papers, and for improperly ordering cars within the train containing hazardous materials (*In re* Union Pacific Railroad, FRA Case No. UP-86-12).

Southern Pacific Railroad, a carrier, paid a $4,000 penalty for failing to provide a train crew with documents showing the position in the train of loaded and placarded tank cars containing potassium hydroxide (*In re* Southern Pacific Railroad, FRA Case No. SP-84-6).

Federal Aviation Administration

Federal Express, an air carrier, paid a $3,000 fine for accepting a shipment of gasoline. The shipment was not properly marked and not properly secured in the aircraft (*In re* Federal Express, FAA Case No. 87WP750045).

Chevron U.S.A., Inc., a shipper, paid a $40,000 penalty for improperly marking, labeling, and packaging an air shipment (chemicals were not properly identified). During the flight, the pilot experienced physical discomfort as a result of the fumes emanating from a leaking package in the shipment (*In re* Chevron U.S.A., Inc., FAA Case No. 86WP750127).

Kerry A. Koehler, a shipper, paid a $550 fine for transporting a tote bag that contained a wood hearth block with approximately two bundles of matches that ignited and started a fire (matches are considered a hazardous material). The shipment was improperly classed, described, packaged, marked, and labeled (*In re* Koehler, FAA Case No. 86CE720011).

Federal Highway Administration

Bell County Oil & Supply Inc., a shipper, paid a $9,000 fine for failing to maintain a driver qualification file for each employed driver, failing to require a driver to make and submit a record of duty status, and transporting hazardous materials with improperly prepared shipping papers.

Fleet Transit Inc., a carrier/shipper, paid a $3,000 penalty for failing to file a written report of an accident involving hazardous materials.

Keystone Fireworks Mtg. paid a $3,000 fine to settle two counts of failing to maintain minimum levels of fiancial responsibility.

U.S. Coast Guard

Chartering Marine Service paid a $1,700 fine to settle charges that the vessel's dangerous cargo manifest did not contain the vessel's official number or call sign, the master's signature, the signature of the preparer and date, the proper shipping name, and the gross weight of each package (*In re* Chartering Marine Service, USCG Case No. MV88001667).

Kwongyuen Hangkee Co. Ltd. paid a $5,000 penalty for accepting a container for transport that was incorrectly placarded. Not all packages in the container were blocked and braced to prevent movement. Some of the boxes containing fireworks were not properly labeled and were missing the required marking (*In re* Kwongyuen Hangkee Co. Ltd., USCG Case No. MV88003074).

Kerr Steamship Co. paid a $175 fine for an improperly completed dangerous cargo manifest, which did not include the classification and stowage location of each hazardous material (*In re* Kerr Steamship Co., USCG Case No. MV87006147).

RSPA

Bonkosky Transportation Inc., a carrier, paid a $1,250 penalty for failing to submit a hazardous materials incident form (DOT Form 5800.1) within the required time period (*In re* Bonkosky Transportation Inc., RSPA Case No. 88-52-HMI).

Buddy's Fire Protection Sales & Service, a cylinder retester, paid a $2,750 fine for performing hydrostatic testing with inaccurate equipment, failing to maintain records of reinspection and retesting, and failing to properly mark the identification number (*In re* Buddy's Fire Protection Sales & Service, RSPA Case No. 88-23-CR).

Van Leer Containers Inc., a steel drum manufacturer, paid a $4,500 fine for making, marking, and selling containers without performing the required drop test, hydrostatic test, and leakage test (*In re* Van Leer Containers Inc., RSPA Case No. 88-07-DM).

Examples of Federal Court Proceedings

In a civil case, an Ohio-based company was charged with failing to file a hazardous materials incident report. DOT levied a $1,000 penalty. The company appealed the amount of the penalty to the U.S. district court. The court upheld the penalty (*People Cartage Inc. v. DOT*, No. 89-CV-1754, (N.D. Ohio Sept. 14, 1991)).

In a criminal case, an official from a Georgia-based cargo tank maker, Georgia Equipment Manufacturer, pleaded guilty to making tanks that did not meet HMTA specifications. The allegations included failure to comply with the terms of an exemption (U.S. Bratschitsch, CR 86-05N, 1987).

II. IDENTIFICATION AND TRANSPORT OF WASTE UNDER RCRA

INTRODUCTION

The Resource Conservation and Recovery Act (RCRA) was enacted in 1976 with an emphasis on control of solid waste and creation of incentives for recylcing. By the time of its amendment in 1984, hazardous waste had become the focus, particularly as a result of the well-known problems with the Superfund hazardous waste site cleanup program: large expenditures of money, along with very few completed cleanups and many questions about the health and environmental protection afforded by the program.

The Superfund program—the Comprehensive Environmental Response, Compensation, and Liability Act (CERCLA)—was primarily directed at cleaning up abandoned hazardous waste sites, but had fallen prey to mismanagement during the early 1980s. Continuing problems with hazardous waste handling and improper disposal motivated Congress to include a series a very strict requirements for ongoing waste operations in the 1984 RCRA amendments.

RCRA is again facing amendment, perhaps within the next year. Ironically, the focus will probably return to the issue of managing solid waste. Problems with disposal are running in two directions: federal and state regulators now understand more about the threats posed by waste disposal, at the same time that the country is running out of landfill space. The problem is compounded by the hostile attitude of most local governments toward siting of new landfills—a manifestation of the "not in my backyard" (NIMBY) syndrome.

For the time being, most of the activity taking place at the Environmental Protection Agency (EPA) under RCRA centers on preparing standards for the proper disposal of hazardous waste. The first step is to determine whether a substance is a "solid waste." If so, that waste may also qualify for stiffer regulation as a "hazardous waste."

Generators of hazardous waste are subject to RCRA limitations and are responsible for initiating the tracking system through the use of manifests that follow the waste through its travels, the basis for the "cradle-to-grave" regulatory system Congress envisioned for the law. By tracking the movement of waste, Congress wanted to ensure proper management along with providing EPA a way to get a handle on volume, placement, and movement. In addition, generators are allowed to accumulate waste on-site for only a limited period of time before they trigger full waste disposal permitting requirements.

A new group of generators, those handling small quantities of waste, were brought under the regulatory scheme in the 1984 amendments. Congress balanced its concerns that wastes from this group could still harm the environment against fears that over-regulation could be an economic burden. This group includes generators who produce from 100 to 1,000 kilograms (220 to 2,200 pounds) of waste a month, and they are held to less restrictive standards than generators of larger amounts of hazardous waste.

IDENTIFICATION, MANAGEMENT OF SOLID WASTE

The first step in determining whether a substance is subject to regulations under RCRA is to figure out whether the substance is a "solid waste" under the law—which is not limited to solid substances. If it is a solid waste, the generator must take the next step of determining whether it also qualifies as a hazardous waste.

Solid Waste

Congress defined solid waste in RCRA as garbage, refuse, sludge, "and other discarded material, including solid, liquid, semisolid or contained gaseous material resulting from industrial, commercial, mining, and agricultural operations" (42 U.S.C. 6903(27)). Note that the definition is not limited to materials that are only in a solid form. However, the term does not include "solid or dissolved material in domestic sewage, or solid or dissolved materials in irrigation return flows"; these and other kinds of material are regulated under the Clean Water Act.

EPA defines solid waste by asking first whether the substance is a discarded material. One group of materials automatically considered to be solid waste includes discarded material: (1) garbage, such as milk cartons and coffee grounds; (2) refuse, such as metal scrap, empty containers, and wall board; and (3) sludges from waste treatment plants. Discarded material also includes substances that are abandoned through disposal or incineration, recycled, or considered to be inherently waste-like (40 CFR 261.2(a)). The agency goes on to identify groups of wastes that are considered inherently waste-like, even when they are recycled, such as wastes from the production of trichlorophenol, pentachlorophenol, or hexachlorobenzene.

Exclusions

There are also materials that are not considered by EPA to be solid wastes as long as they are recycled (40 CFR 61.2(e)). These include:

- Materials that are used or reused as ingredients in an industrial process to make a product.
- Materials used as substitutes for commerical products.
- Materials returned to the original process from which they are generated, as long as they are not first reclaimed. EPA also requires that the material be returned as a substitute for a raw material. If the material is reclaimed and returned to the original production process, the agency imposes a separate set of requirements if the waste generator wants to avoid having the material designated as solid waste (40 CFR 261.4(a)(8)).

Paralleling the definition under RCRA, EPA's regulations under the law hold that domestic sewage and discharges permitted under the Clean Water Act are not considered solid waste (40 CFR 261.4(a)). Other listed exclusions include special nuclear materials defined under the Atomic Energy Act, pulping liquors, and spent sulfuric acid used to produce virgin sulfuric acid.

Nonhazardous Solid Waste

Another group of materials has been designated as solid waste but not taken to the next level of regulation as hazardous waste. This list is mostly limited to specific wastes produced by designated industries, such as certain wastewater sludges produced by the leather tanning and finishing industry (40 CFR 261.4(b)(6)). Other examples of nonhazardous wastes include:

- Household wastes, which include garbage, trash, and sanitary wastes, and the ash produced from the burning of such wastes by municipal incinerators
- Solid wastes created by growing crops or raising livestock, which are then used as fertilizer
- Mining waste from the extraction and processing of ores and minerals
- Cement kiln dust

State Plans

Once material is designated as a nonhazardous solid waste, the requirements for handling the material are only broadly indicated in the federal law. Under RCRA Subtitle D, the responsibility for regulating solid waste falls on state authorities (42 U.S.C. 6946). The only clear federal requirement is that solid waste not be disposed of at "open dumps." In turn, states have developed requirements for different types of solid waste facilities, with some states imposing requirements for landfills that are comparable to those imposed under RCRA for hazardous waste facilities.

EPA has been working on standards for municipal solid waste landfills, which were first proposed in 1988. After a lengthy period of review by the White House Office of Management and Budget, the rules were reproposed in summer 1991. The key issues in the proposal include the extent of groundwater monitoring, use of liners to trap leachate and prevent leakage, and the costs associated with increased protection.

For now, the federal government provides minimum criteria for states to use in regulating landfills and other solid waste facilities. State plans must:

- Identify the responsibilities of state, local, and regional authorities in carrying out a solid waste plan
- Create a regulatory scheme for identifying and closing or upgrading open dumps, as well as for prohibiting the creation of new open dumps

- Ensure that adequate capacity exists to dispose of solid waste properly
- Ensure public participation in development of plans, regulations, and facility permitting

States must also consider the effects of solid waste disposal facilities and open dumps, with an emphasis on floodplains, endangered species, surface water, groundwater, land application, disease, air, and safety (40 CFR Part 257). In addition, the United States provides grant money for state and local authorities to develop solid waste management plans (42 U.S.C. 6948).

IDENTIFICATION OF HAZARDOUS WASTE

Once a waste generator determines that a material is a solid waste, the next question is whether it is also a hazardous waste. The definition of hazardous waste in the statute is descriptive as to the threat posed by such waste, but is not specific (42 U.S.C. 1004(5)). Hazardous waste is a solid waste that may:

> (A) cause, or significantly contribute to an increase in mortality or an increase in serious irreversible, or incapacitating reversible, illness; or
>
> (B) pose a substantial present or potential hazard to human health or the environment when improperly treated, stored, transported, or disposed of, or otherwise managed.

Congress fleshed out the specifics a bit more in another provision of RCRA, setting timetables for EPA to consider whether various groups of wastes qualify as hazardous. The agency has carried out these requirements in its listings of hazardous wastes. Waste generators must be aware of the general characteristics that EPA looks to in determining whether a substance is hazardous. At the same time, any given substance may be specified as hazardous on the lists contained in EPA's regulations.

Hazardous Waste Characteristics

Part of the task in identifying hazardous waste is to determine whether it exhibits any of the following characteristics (40 CFR 261, Subpart C):

- *Ignitability:* wastes that have a low flash point or that spontaneously combust and burn.
- *Corrosivity:* wastes that are extremely acidic or basic (with a high or low pH) or that can corrode steel at a rapid rate.
- *Reactivity:* wastes that react violently with water or are easily exploded—what EPA has called "unstable" wastes, such as water from TNT operations and used cyanide solvents.
- *Toxicity characteristic leaching procedure:* a method of determining whether a waste is likely to leach toxic material if disposed of improperly. The test attempts to measure the potential contamination by simulating leaching conditions at a landfill and determining the types and concentrations of toxic constituents that result. Thirty-nine types of organic chemicals are tested.

Hazardous Waste Lists

EPA has provided lengthy lists of wastes that are automatically considered to be hazardous. The agency decides to list a waste if it exhibits toxic characteristics. A waste generator simply has to check these lists to determine whether its waste is subject to regulation under RCRA as hazardous. EPA has identified three groups of listed wastes: hazardous waste from nonspecific sources, such as wastewater sludges from electroplating operations and some types of solvents from degreasing operations (40 CFR 261.31); hazardous waste from specific sources, such as distillation residues from production of certain types of chemicals or various kinds of dust and sludge from emission control devies (40 CFR 261.32); and discarded commercial chemical products, spill residues, and other substances occurring in commercial operations (40 CFR 261.33).

Delisting

A waste generator can petition EPA to remove a substance from the lists if the generator feels the waste does not qualify for or require such stringent regulation; however, the agency places a substantial burden on the petitioner (40 CFR 260.22). The generator must show that the waste produced by that particular facility does not have the characteristic that EPA thought merited the listing. In addition, the generator must show that the waste does not exhibit other hazardous characteristics. It should also be noted that there have been long-standing complaints about the length of time EPA takes to process delisting petitions: in the 1984 RCRA amendments, Congress responded by imposing a two-year deadline on EPA to issue a final decision on a petition.

One last group of wastes is considered hazardous by EPA: some mixtures of solid and hazardous wastes (40 CFR 261.3). Generally, any mixture containing a listed hazardous waste must be managed as hazardous. However, for wastes that were listed based on a toxic characteristic, the agency will be willing to look at whether the mixture retains that characteristic.

REGULATION OF GENERATORS

The RCRA system for tracking the movement of hazardous waste, a "cradle-to-grave" system, is triggered by the actions of persons who generate such waste (40 CFR Part 262). The generator has the responsibility to first determine that it is creating hazardous waste. Once that determination is made, the waste generator is required to contact a regional EPA office to obtain an identification number for the waste operation. If a generator fails to identify material as hazardous waste, and fails to comply with RCRA requirements, state or federal regulators may later impose heavy civil or criminal penalties.

The tracking system begins with the generator preparing a manifest that will travel with the waste through its transport and final disposal. The Uniform Hazardous Waste Manifest requests information about types and amounts of waste, the containers used, special handling instructions, and information about the designated waste transport companies and disposal facilities. Copies of manifests must be retained by the generator for three years.

Generators may accumulate waste on-site for up to 90 days but must provide proper storage. If wastes are accumulated for a longer period, the generator must obtain permission from the regional EPA office, which will consider requests on a case-by-case basis; retention of waste for such a period will subject the generator to the more egregious requirements imposed on storage facilities.

Other reporting requirements are also imposed by EPA. Generators must submit biennial reports on their activities to the agency's regional offices. In addition, generators must notify EPA if they do not receive confirmation of the waste having reached its destination.

Small-Quantity Generators

A new category of generators was created in the 1986 amendments to include those that generate small quantities of hazardous waste (100 to 1,000 kilograms per month) and transport the waste off-site. In general, this group is required to follow the same rules as all others:

- Obtain an EPA identification number
- Properly handle the waste before transport
- Make sure the manifest accurately reflects the waste and follows it

- Fulfill recordkeeping and reporting requirements

However, small-quantity generators may accumulate waste on-site for 180 or 270 days, depending on the distance the waste will be transported (40 CFR 262.34(d)–(f)). In addition, recordkeeping requirements are slightly less onerous than for other generators—with an exemption from biennial reports (40 CFR 262.44).

HAZARDOUS WASTE TRANSPORT

In preparing waste for shipment, the generator must comply with DOT's requirements for packaging (49 CFR Parts 173, 178, 179). In addition, waste containers must be marked with appropriate identification and warnings concerning their contents. The DOT regulations specify requirements for packaging material, inspection, and testing of containers for hundreds of types of hazardous substances. In addition, they include specifications for shipping containers ranging from metal barrels to portable tanks and tank cars.

Persons engaged in the transport of hazardous waste must meet a separate set of EPA requirements, although the regulations are relatively limited (40 CFR 263). The limited exception to the transport requirements involves a waste generator moving waste on-site. On the other hand, a transporter can trigger the full RCRA requirements for a waste generator by importing hazardous waste into the United States or by mixing different wastes in a single container. Or a transporter can become subject to extensive RCRA storage requirements by storing wastes for longer than 10 days.

A waste transporter has three primary responsibilities:

- Obtain an EPA identification number.
- Comply with the waste manifest and recordkeeping system. This duty is the most extensive under the regula-

tions, but consists of simply ensuring that the waste manifest is properly signed and dated at the point where the waste is picked up and later transferred and that the manifest is accurate. The transporter retains a copy of the manifest for three years.

- Take immediate action if there is an accidental discharge of waste. If there is a discharge, the transporter must notify government authorities (federal, state, and local) and take response actions as approved by these authorities.

Chapter 5

STORAGE

I. REGULATION OF STORAGE TANKS UNDER RCRA

INTRODUCTION

One of the new areas of regulation in the 1984 amendments to the Resource Conservation and Recovery Act (RCRA) involved underground storage tanks (UST). In an entire subchapter of the law, Congress demanded that the Environmental Protection Agency (EPA) and state authorities take steps to address the contamination potential posed by the estimated 3 million underground tanks in the United States (42 U.S.C. 9001, *et seq.*). Most of these tanks store petroleum products; less than 3 percent contain chemicals listed under the Superfund law.

EPA and petroleum industry experts have estimated that as many as 20 percent of these tanks are leaking, causing substantial contamination of groundwater and drinking water sources. The reason most cited is that the majority of older storage tanks were built with bare steel and are subject to corrosion.

Under the UST program, tank owners must notify state authorities about the existence and condition of their tanks, and tank owners and operators must demonstrate compliance with RCRA requirements. EPA has issued regulations on design and construction standards, proof of financial responsibility by tank owners and operators, and deadlines for compliance. In addition, states have authority to carry out programs that have been approved by the agency.

PROGRAM SCOPE

Only about one third of U.S. underground tanks are covered in the RCRA program, according to EPA; the remainder are covered by other programs or do not pose a hazard. Of the group that is regulated, the agency estimated that 676,000 tanks are used for retail motor fuels, 651,000 are used for petroleum storage, and 54,000 store chemicals.

EPA defines an underground tank as any tank with 10 percent or more of its volume buried underground, including any pipes that are attached. Tanks subject to the RCRA program include those storing petroleum products or any substance defined as hazardous under the Comprehensive Environmental Response, Compensation, and Liability Act (CERCLA) (40 CFR 280.12).

Several types of tanks are excluded from regulation (40 CFR 280.10, 280.12), such as:

- Tanks used to store hazardous wastes regulated under RCRA Subtitle C
- Tanks with a capacity of less than 110 gallons
- Wastewater treatment tanks used as part of a wastewater treatment facility regulated under the Clean Water Act
- Farm or residential tanks holding 1,100 gallons or less of motor fuel for noncommercial purposes
- Tanks used for storing heating oil that will be used on the premises where stored

- Tanks on or above the floors of underground areas, such as basements or mining shafts

Owners and operators of regulated tanks had until May 1986 to notify state and local authorities about each tank's age, size, type, location, and use. For any tank that first came into use after May 1986, notice must be provided to the appropriate state or local agency within 30 days (40 CFR 280.22). (See also Appendices I and II of 40 CFR 280 for a copy of the notification form and listing of state and local agencies.)

TECHNICAL REQUIREMENTS

The technical requirements address the common tank problems of corrosion and the resulting leakage that can occur. Releases must be reported, and owners may then be required to perform corrective action. Also, proper measures must be taken when closing down the tanks.

Alternatives for addressing corrosion control include (40 CFR 280.20(a)):

- Building tanks with fiberglass-reinforced plastic
- Using a steel tank that is cathodically protected
- Constructing tanks with a steel-fiberglass-reinforced-plastic composite
- Using a metal tank at a site where corrosion is not likely to be a problem during the life of the tank
- Using a manner of tank construction and corrosion protection that regulatory agencies consider to be as protective as the above methods

Tank owners also must install equipment to prevent spills and overfilling (40 CFR 280.20(c)). For example, a spill catchment basin can be built to prevent releases when the hose for filling the tank is removed. To preclude overfilling, a mechanism must be in place to shut off flow when the tank is 95 percent filled or to sound an alarm or restrict flow when the tank is 90 percent filled. These requirements may be bypassed when the owner provides an equally protective alternative or when less than 25 gallons will enter the tank at any one time.

Existing tanks must be upgraded; owners have until December 22, 1998, to do so. The upgrade must meet the same performance standards as new tanks or follow EPA requirements on performance standards for interior lining, cathodic protection, and spill and overflow prevention (40 CFR 280.21).

Further technical requirements address installation of leak detection systems at tank sites. The various methods allowed by EPA include (40 CFR 280.43):

- Monthly inventories of products, with extreme accuracy
- Use of manual gauges (similar to "dipsticks")
- Tank tightness testing
- Vapor monitoring, if soils where the tank was placed allow it
- Groundwater monitoring
- Interstitial monitoring between the tank and a barrier around it

Deadlines for compliance with existing tanks depend on the age of the tank—with a beginning point of December 1989 for tanks installed before 1965.

If a release is detected, the owner or operator has 24 hours to report it to the proper authority (40 CFR 280.50). Reporting is triggered (1) by a release of 25 gallons or more of a petroleum substance, (2) by a release of a reportable quantity of a CERCLA hazardous substance, or (3) where a release of a lesser amount cannot be cleaned up within 24 hours. Once reported, corrective action will be required to clean up the release (40 CFR 280.60).

At the end of the tank's life, the owner or operator must notify the regulatory agency, empty the tank, and remove it or fill it with an inert solid substance such as sand (40 CFR 280.71).

FINANCIAL REQUIREMENTS

This aspect of the tank regulations probably represents the greatest burden, particularly to owners and operators of small numbers of tanks. Congress intended that the industry demonstrate financial responsibility to ensure proper tank closure and pay for bodily injury or property damage. However, as with insurance for hazardous waste facilities, tank owners and operators are having problems finding insurers or insuring themselves.

The breakdown of coverage requirements is somewhat complicated (40 CFR 280.93). For underground tanks used to store petroleum products, the required coverage is:

- $1 million per occurrence, or single leak, for tanks at petroleum marketing facilities that handle an average of more than 10,000 gallons per month based on annual throughput for the previous calendar year
- $500,000 per occurrence for all other types of petroleum facilities
- $1 million to cover leaks that might occur over a year if there are one to 100 tanks at the facility
- $2 million annual aggregate if there are 101 or more tanks at the facility

For underground tanks not used at petroleum facilities, the required coverage is:

- $1 million per occurrence for facilities that have a monthly throughput of more than 10,000 gallons
- $500,000 per occurrence for facilities with less than 10,000 gallons of monthly throughput
- $1 million to $2 million of annual aggregate coverage, depending on the number of tanks at the facility

The complexity of the deadlines for compliance depends on the type of facility at issue, the number of tanks, and the net worth of the facility owner or operator. The deadlines range from January 24, 1989, to April 1991.

STATE UST PROGRAMS

Many states had underground tank programs in operation before the federal government took action. The federal program began with the 1984 RCRA amendments; as with other parts of RCRA, states still must be delegated the authority to run programs that are in line with federal requirements. At present, almost all states have formal agreements in place to carry out UST activities, although none have yet been delegated formal program authority. EPA's other roles have been to assist states in developing tank insurance funds and to encourage them to develop financial assistance programs to help small businesses comply with UST requirements.

The agency has issued regulations for approving state UST programs, which have the same elements as contained in the federal program (40 CFR 281). For instance, state tank programs must cover (1) design, construction, installation, and notification for new UST systems; (2) upgrading of existing tank systems; (3) release detection and reporting mechanisms; (4) financial responsibility guarantees; and (5) proper closure. In addition, the state must be able to show that it can enforce program requirements.

State program requirements must be at least as stringent as those contained in the federal program. However, a state program that is less stringent can be granted interim approval, which is good for only a limited period of one to three years; the state must still submit an amended program to EPA for approval. EPA's authority does not end once the state program is approved: the agency retains the right to monitor the state's progress and enforce program requirements on its own.

OTHER PROVISIONS

Trust Fund

In response to potential shortages in financing for cleanup of underground tank contamination, Congress created a Leaking Underground Tank Storage Tank Trust Fund. The fund is financed by a tax on gasoline, diesel fuel, and aviation fuels, and is to be used when no private party is willing or able to pay for cleanup of petroleum contamination (42 U.S.C 6991(b)(h)). The federal or state government is entitled to sue responsible parties to recover costs, similar to the system under CERCLA.

Enforcement

The first step is for EPA to issue a compliance order for any violation of RCRA tank requirements or the agency's regulations (42 U.S.C. 6991(e)). If the alleged violator fails to comply with the order, EPA can seek a civil penalty of $25,000 per day of noncompliance. EPA can also opt to forgo the compliance order and seek a civil penalty of up to $10,000 per tank per day. There are no criminal penalty provisions under this portion of RCRA.

II. HAZARDOUS WASTE STORAGE UNDER RCRA

Storage of hazardous wastes generally subjects a facility to all the requirements of RCRA for "treatment, storage, and disposal facilities." Storage units are usually defined as tanks, containers, waste piles, and certain other impoundments. Storage requirements also come into play when wastes are removed from a facility during closure.

Section 261.4(c) excludes from hazardous waste permit requirements the storage of a waste for less than 90 days. This covers wastes in "a product or raw material storage tank, a product or raw material pipeline, or in a manufacturing process unit or an associated non-waste-treatment-manufacturing unit." In addition, RCRA interim status permit requirements do not apply to transporters that store manifested hazardous waste at a transfer facility for 10 days or less (Section 265.1(c)(12)).

Facilities used for recycling wastes are exempt from RCRA storage requirements; however, if a recyclable waste is used in a way that constitutes disposal, storage requirements would be imposed just as other RCRA standards would be.

Solid waste storage standards are less onerous than those for hazardous wastes. Section 243.200 of the RCRA rules simply requires that solid waste storage not create a fire, health, or safety hazard and that it not harbor pests that carry diseases. Containers must be leakproof, durable, nonabsorbent, and capable of containing all waste generated between collections.

III. HAZARDOUS MATERIAL STORAGE UNDER TSCA

The Toxic Substances Control Act (TSCA) addresses in some detail the manufacture and distribution of chemicals, but it says little about storage. Even though storage generally lies between those two activities, the Environmental Protection Agency (EPA) retains broad authority to act on imminent hazards during storage. The agency can inspect any chemical storage facility and subpoena records under TSCA Section 11.

TSCA also has chemical-specific restrictions, some of which extend to storage. The regulations on polychlorinated biphenyls (PCBs) contain the most extensive storage requirements.

TSCA's PCB disposal requirements depend on the concentration of the substance. PCB leaks, spills, or other uncontrolled discharges are considered improper disposal and may be subject to civil charges. To ensure that PCB disposal rules are observed, each EPA regional office has a person designated as the PCB disposal contact. The agency also maintains a list of companies it has issued PCB disposal permits under TSCA. Companies that do not meet the terms of their PCB storage and disposal permits may face revocation.

PCBs in concentrations of between 50 parts per million (ppm) and 500 ppm generally must be disposed of by incineration in a TSCA-permitted facility, in a chemical waste landfill, in a high-efficiency boiler, or in an EPA-approved alternative combustion process. Items that are regulated by the PCB disposal rules include:

- Mineral oil dielectric fluid
- PCB-containing liquids
- Contaminated rags, soil, and other debris
- Dredged materials and sludge
- PCB articles, transformers, capacitors, hydraulic machines, contaminated electrical equipment, and containers

PCBs placed in storage must be disposed of within one year. Storage facilities must meet certain requirements, such as having an adequate floor (including a minimum six-inch-high curb and no floor drains). Some PCB-containing items may be stored in an area that does not meet the storage facility requirements, but such items cannot be stored at a temporary facility for more than 30 days. In addition, records must be maintained on the amounts of PCBs or PCB items stored, received, or transferred during the year.

Commercial PCB storage facilities that store more than 500 liquid gallons of PCB waste must meet EPA standards and be certified by the agency. In addition to documenting the facility's ability to handle the PCB material, the owner/operator must have an acceptable closure plan, estimate closure costs, and demonstrate an ability to meet the financial obligations of closing the facility. Storage areas at facilities permitted under the Resource Conservation and Recovery Act (RCRA) may be exempt from these TSCA requirements.

EPA established a cradle-to-grave tracking system for PCB wastes modeled on the RCRA hazardous waste tracking system. It requires companies involved in disposing of PCB wastes to notify the agency and obtain identification numbers. Only generators of PCB wastes with concentrations of more than 50 ppm must prepare TSCA manifests using the Notification of PCB Activity form (EPA Form 7710-53).

Each party that handles the PCB waste must sign and retain a copy and forward the remaining copies of the manifest with the waste until it reaches the storage or disposal facility. If a discrepancy is discovered between the type and quantity of waste described on the manifest and the material received, the operator of the facility must reconcile the problem with the transporter or generator. If the problem cannot be resolved within 15 days, the operator of the storage or disposal facility must notify EPA.

The PCB waste generator must keep a copy of the manifest until a signed copy from the facility designated to receive the wastes is received. This signed, complete manifest must be retained for at least three years. The manifest used for tracking PCB wastes is the same one used for tracking RCRA hazardous wastes (EPA Form 8700-22). Manifest reporting requirements under RCRA are discussed in detail in Chapter 1.

The owner/operator of a disposal facility must prepare a certificate of disposal for each shipment of manifested PCB waste. This document summarizes the data from the manifest and certifies the disposal of the identified wastes. Generators must retain their copy of the disposal certificate for three years.

IV. PESTICIDE STORAGE UNDER FIFRA

The Federal Insecticide, Fungicide, and Rodenticide Act (FIFRA) requires that pesticide registrants provide EPA with information on safe storage and disposal methods. This information may be required under Section 3 (product registration) or Section 6 (registration cancellation and suspension). FIFRA Section 19 includes requirements for pesticide storage and container design. Most registered pesticides have storage and disposal instructions on their labels.

EPA published recommendations for safe storage of pesticides in 40 CFR 165. In general, pesticides must be stored in appropriate facilities, and all dangerous pesticides should be labeled to alert people of their hazards.

Temporary storage sites are allowed when a pesticide will be used immediately; however, only the amount intended for application at the site may be stored there. Such sites must be secure from flooding, entry by unauthorized persons, and other potential causes of release.

Long-term storage facilities should be separate rooms, buildings, or covered areas that are dry, ventilated, and protected from fire. The facilities must be secure from unauthorized entry and, if necessary, surrounded by a climb-proof fence. Equipment that comes in contact with pesticides should bear warnings of contamination if it can be used for other purposes.

Storage facilities should have some provision for decontamination of people working with pesticides. Containers should be checked regularly for signs of corrosion or leakage. Operators of storage facilities also should adopt safety measures and accident prevention plans to prevent exposure of workers and nearby residents.

V. EPCRA NOTIFICATION REQUIREMENTS

The Emergency Planning and Community Right-to-Know Act (EPCRA) requires companies to notify state authorities when they possess any "extremely hazardous substances" above threshold quantities, and this extends to storage. About 300 chemicals now are listed as extremely hazardous under EPCRA, and reporting thresholds vary from one pound to 10,000 pounds. Additional information on EPCRA reporting may be found in Chapter 2.

Chapter 6

DISPOSAL & CLEANUP

I. DISPOSAL OF WASTE UNDER RCRA

INTRODUCTION

The key statute addressing the handling and disposal of hazardous waste is the Resource Conservation and Recovery Act (RCRA). In particular, RCRA established the permit process for facilities that handle hazardous waste. Before the law was amended in 1984, controls on waste treatment, storage, and disposal facilities were relatively lax, and supervision by the Environmental Protection Agency (EPA) and state governments was minimal. Congress responded by imposing a set of stringent technical requirements that have led to a revolution, and some panic, in the waste industry.

In addition to providing a liner or buffer between the waste disposal unit and the environment, facility operators must demonstrate that they are monitoring groundwater in the area for signs of contamination. Another requirement has proven especially difficult for facility owners: they must show that they have insurance to cover potential problems.

At present, however, very few insurers are willing to provide such coverage at reasonable prices.

A major regulatory effort to ban the land disposal of hundreds of untreated hazardous wastes is under way, proceeding in several stages. Various rounds of regulation have addressed treatment of solvents and dioxins, a group of wastes also known as the "California List," and EPA has just completed the "Thirds" group of hazardous wastes. Industry can expect the annual cost of complying with these new standards to range in the billions of dollars. At the same time, facilities able to upgrade to the treatment technologies mandated by these requirements can expect increased business; those that fail technological requirements will shut down.

TREATMENT, STORAGE, AND DISPOSAL FACILITIES

The most pervasive requirements under RCRA apply to operators of facilities that handle hazardous waste. These treatment, storage, and disposal (TSD) facilities must handle wastes in a specified manner, comply with technology and financial assurance requirements, prepare an adequate closure plan, and undergo inspection at regular intervals. In the 1984 amendments to RCRA, Congress paid particular attention to hazardous waste facilities in the belief that strict regulation of their operation would prevent them from becoming abandoned Superfund sites in the future.

The first issue to address is whether an industrial site qualifies as a TSD facility. If it does, the facility operator must have a final (also called a "Part B") operating permit or must have submitted a proper permit application. These permit requirements will be triggered for a range of facilities, including industrial areas where the production of wastes can lead to full TSD regulation for the unwary operator. Permit requirements are stringent and cover factors such as groundwater monitoring, types of waste the facility is authorized to handle, technology required for waste-holding units, and corrective action to remedy releases of waste.

What Is a TSD Facility?

At the most basic level, RCRA standards apply to any facility that treats, stores, or disposes of hazardous waste and that existed or was under construction as of November 19, 1980. This generic definition is expanded somewhat by EPA's own definitions (40 CFR 260.1(a)).

A "facility" includes land and structures where hazardous waste is handled. Each part of the facility is identified as a "hazardous waste management unit," which the agency identifies as "a contiguous area of land on or in which hazardous waste is placed," such as a waste pile, landfill cell, or waste tank. Types of facilities covered include:

- Treatment facilities, which use any method designed to neutralize or change the composition of waste so as to make it easier to handle, including reducing the volume of the waste
- Storage facilities, which hold waste for a temporary time, in preparation for later handling
- Disposal facilities, where "hazardous waste is intentionally placed into or on any land or water" and the waste remains there after facility closure

While the coverage of the definitions is extensive, EPA has provided some exemptions, including (40 CFR 264.1):

- Facilities that handle waste in compliance with a state program that has been approved by EPA (although state program requirements will basically be the same as, or even more stringent than, national requirements)
- Facilities that manage recyclable materials
- Generators and transporters who store waste for only a temporary period

- Wastewater treatment facilities or public sewage treatment systems subject to Clean Water Act regulation
- Persons engaged in emergency response to the discharge or threatened discharge of a hazardous waste

Limited exemptions are also available when hazardous waste is dumped in the ocean subject to a permit under the Marine Protection, Research, and Sanctuaries Act or injected underground subject to the Safe Drinking Water Act.

Interim Status Facilities

Before addressing requirements for fully permitted facilities, it must be noted that EPA has prepared a set of requirements for facilities still awaiting permit approval (40 CFR Part 265). These waste-disposal facilities were required to submit an application for a final permit by November 8, 1985, along with certifying compliance with standards for groundwater monitoring and financial assurance. At that time, about 1,600 facilities were subject to the permitting requirements.

In effect, there is little difference between the requirements imposed on interim status and Part B permit facilities: both must meet general standards for technology, monitoring, proof of financial capability, and site security. The interim status requirements allow facilities with a waste management system in place to continue to operate while the application is evaluated, although the state or federal regulatory agency can certainly demand further modifications or even closure. EPA has said that the somewhat less stringent requirements for interim status facilities are "good housekeeping" standards. Even though much of the focus on facility permits is on final operating standards, interim status requirements are still an essential part of EPA's regulatory system under RCRA.

However, failure to obtain even interim status approval means that a facility must close. It is estimated that about one third of the hazardous waste operations in the United

States had to close as of the November 8, 1985 deadline because the requirements that went into effect at that point were so stringent. At the time of the deadline, only 492 facility operators had actually certified that they met final RCRA permit requirements. EPA recognized the hardship imposed by the deadline, especially the financial responsibility requirements, and offered to consider the individual circumstances of facility operators who were still attempting to obtain final approval.

Interim status issues frequently arise in EPA proceedings to enforce RCRA, with resulting disputes over whether the facility operator complied with permit application and certification requirements by the November 1985 deadline. For example, in *U.S. v. Allegan Metal Finishing Co.* a metal finishing company was held liable for violating RCRA even though the company asserted that it was impossible to meet EPA's financial requirements by the statutory deadline (696 F. Supp. 275, 28 ER Cases 1581, (W.D. Mich. 1988)).

Final Permits

At the heart of the 1984 RCRA amendments were the new standards applied to TSD facilities in final permits. These standards are comprehensive, stringent, and relatively costly to implement. The goal was to prevent such facilities from releasing hazardous waste or becoming future candidates for cleanup under the Superfund law.

Sections 3004 and 3005 of RCRA are unusually specific in the requirements EPA must incorporate in issuing final permits to TSD facilities (42 U.S.C. 6924, 6925). Congress went into great detail and imposed deadlines for regulatory action, based on concerns in the early 1980s that EPA was compromising environmental protection in the face of political and economic pressures. The agency codified these requirements in its own regulations for permitted facilities (40 CFR Part 264).

A series of extensive requirements were imposed on facility owners and operators:

- *Administrative Requirements:* Facility operators must obtain an EPA identification number, provide an analysis of wastes to be handled at the facility, lay out a schedule for self-inspection of equipment, install a security system, and ensure proper training of personnel at the site. In addition, facility operators must sign the manifest that follows a hazardous waste and provide other copies to the waste transporter and generator.

- *Technical Requirements:* Operators of TSD facilities must have in place a series of wells to monitor groundwater for contamination. The wells must be placed both at points before groundwater flows under the facility (up gradient) and after it passes under the facility (down gradient). The resulting water samples must be checked for chemical constituents that are handled by the facility. New and modified land disposal units must have a double-liner system installed to prevent percolation of residues under the waste unit into groundwater. In addition, a collection system must be installed to collect leachate, or runoff from leakage and rain. In turn, that leachate is subject to treatment. Although some older units were allowed to operate with single-liner systems, Congress has required that such units be retrofitted to meet the new standards.

- *Closure Plans:* These plans must be submitted along with a permit application. They must address how to close individual waste management units, contain an estimated inventory of hazardous wastes at the facility, provide details as to how to remove or decontaminate waste residues, ensure continued monitoring and control of site runoff, and set schedules for closure. EPA's regulations also specify timetables for postclosure activity, with waste disposal facilities required to undergo monitoring for 30 years after closure.

- *Financial Requirements:* These apply to two aspects of facility operation: site closure and liability coverage for accidents. In regard to site closure, operators must provide an estimate of the costs of having a third party close the facility in compliance with EPA requirements. To guarantee the availability of the estimated amount, the facility operator may use a closure trust fund, surety bond, letter of credit, or insurance or can pass a test demonstrating that the operator has adequate finances to complete closure. Similar requirements apply to show that finances are available for taking care of the facility after closure. Liability coverage must be provided to compensate injury and property damage that may arise from facility operations, such as waste discharges. For sudden accidents, coverage must be a minimum of $1 million per occurrence and an annual aggregate of at least $2 million. For nonsudden accidents, the coverage must be $3 million per occurrence with an annual aggregate of at least $6 million. Coverage limits may be combined for sudden and nonsudden accidents, and the operator can obtain coverage through insurance.

One major problem has arisen with the financial assurance requirements: owners and operators have found it almost impossible to obtain the requisite insurance. The insurance industry, fearful of the mounting claims from Superfund cleanups, is leery of dealing with similar liability problems with RCRA facilities. At present, EPA is looking into ways to help by identifying potential insurance sources and examining the financial circumstances pertaining to each facility.

Corrective Action

Requirements for facility operators and owners to respond to waste discharges can pose a major headache in seeking a final permit. Section 3004(u) of RCRA provides that corrective action can be imposed as a permit requirement in response to waste released from a facility, regardless of when the waste was placed in the facility. EPA has prepared fairly terse regulations concerning corrective action (40 CFR 264.100) but has promised for several years to develop more extensive requirements. For the moment, EPA or state permit writers consider corrective action on a case-by-case basis.

In July 1990, EPA issued a proposal that expanded its policy on corrective action at waste facilities (55 FR 39798). This proposal outlined a series of priorities EPA identified for its corrective action program at a possible 4,600 facilities:

- The highest priority will be placed on action at the most environmentally significant facilities and on the most significant problems at specific facilities.
- "Conditional" remedies will be used where prompt remedial action can reduce risk to acceptable levels or where final cleanup is impractical.
- Voluntary cleanups by facility owners and operators are to be encouraged.
- Facility investigations will be focused on plausible concerns and likely remedies, so as to hasten decisions on cleanups.

EPA said that it was not trying to identify specific levels of cleanup for each type of contamination, but retained the view that cleanups are based on a case-by-case approach. This approach is similar to that followed by the agency in the National Contingency Plan under the Superfund law, where specific cleanup standards will be derived from other federal and state laws, depending on the individual site conditions. Industry's concern with corrective action authorities is that they could create liabilities rivaled only by the federal Superfund program: estimates of the potential cost of corrective action regulations have run in the billions of dollars.

There are also questions for the agency to resolve about use of corrective authorities versus use of cleanup powers available in the Comprehensive Environmental Response, Compensation, and Liability Act (CERCLA). Use of CERCLA or RCRA authorities will be based on issues such as the extent that contamination has traveled off-site, the compliance history of the facility, how many and which parties are responsible for the problem, and the ability of the RCRA facility operator to pay for cleanup. EPA has also indicated that abandoned sites will be addressed through CERCLA authorities and operating facilities will be subject to RCRA corrective action.

BANS ON LAND DISPOSAL OF WASTE

Even with the extensive regulation of hazardous waste facilities, Congress felt that there were certain types of hazardous wastes that cannot be disposed of on land in an untreated state. The concern was obvious: hazardous wastes may find their way into surrounding soil and waters, thereby posing an immediate danger to human health and the environment. The categories of waste banned under RCRA include organic solvents, the California List of hazardous wastes, and a set of some 450 other wastes regulated as the Thirds lists of waste (RCRA Section 3003, 42 U.S.C. 6923). For all these types of wastes, treatment will be required prior to disposal on land. Land disposal has been interpreted by EPA to include placement in landfills, surface impoundments, waste piles, injection wells, and land treatment facilities.

- *Organic Solvents and Dioxins:* Under EPA's November 1986 rule covering five solvent- and seven dioxin-containing wastes (40 CFR 268.30, 268.31), most such wastes could be treated through incineration as part of the requirement for "best demonstrated available treatment" (BDAT) prior to land disposal. Use of BDAT should achieve the treatment standard specified by EPA, but the agency said that it accounted for variations in treatment equipment in deriving the standards. The expected cost of compliance to industry is $152 million annually.
- *California List:* In July 1987, the agency issued standards for hazardous wastes with concentrations of polychlorinated biphenyls (PCBs), eight metals, halogenated organic compounds, cyanides, and corrosive wastes that exceed specified levels (40 CFR 268.32). California had first developed this list of wastes and restricted land disposal,

which Congress picked up in the 1984 RCRA amendments. Again, incineration is likely to be the predominant treatment prior to disposal.

- *The Thirds List:* The remaining group of about 450 wastes was divided into three parts for the purpose of issuing regulations. The first-third rule, issued in August 1988, covered 39 wastes (53 FR 31138). The major groups of wastes covered by the rule included organic wastes from petroleum refineries, electroplating sludges, and metal-containing wastes such as those from electric arc furnace dust. Treatment methods included incineration, waste stabilization, and metals recovery, with an expected cost to industry of $950 million annually. In the second-third rule, a subsequent group of 67 wastes was regulated (40 CFR 268.34). Treatment standards were set for cyanide-containing wastes generated primarily by the electroplating industry and the acrylonitrile production industry, with EPA recommending incineration as the standard treatment measure. In the third-third rule, regulations covering nearly 350 remaining hazardous wastes were issued in May 1990 (40 CFR 268.35). Wastes in this round of regulation included those that exhibit hazardous characteristics such as ignitability, corrosivity, reactivity, or capacity to leach toxic constituents to groundwaters. In addition, leachate collected from storage of hazardous waste must be treated along the same lines as the wastes from which the leachate is derived—with the resulting use of a "train" of treatment technologies that would string together the same cleaning technologies used for the underlying wastes.

STATE PROGRAM APPROVAL

As with most other federal environmental laws, states are authorized to carry out their own versions of the law if given approval by EPA (42 U.S.C. 3006). The approval process operates in two stages. First, for states that have submitted programs that are substantially equivalent but have not yet reached the full level of regulatory strength of the federal program, an iterim authorization may be granted. Second, states that can show their waste programs are at least as stringent as EPA's can qualify for final authorization. States are also allowed to have programs that are more stringent than the federal program. At this point, almost all states have final authorization.

The elements of a state hazardous waste program must be the same as described above for all parts of the RCRA hazardous waste control program (40 CFR 271). These elements cover (1) identification and listing of wastes, (2) generators and transporters of wastes, and (3) permitting of waste handling facilities. Also, the state must show that it will be able to enforce all aspects of the program. On submission of materials from the state, EPA will consider whether the state program is equivalent to and consistent with the federal program, whether it is enforceable, and whether it provides for proper public notice before permit issuance.

After approval, further federal action can take place. If EPA revises its RCRA program or regulations, states have up to two years to bring their programs into compliance; however, EPA will retain the right to enforce its requirements in the iterim. If EPA decides in monitoring a state program that the program no longer complies with RCRA requirements, the agency can withdraw approval and take over administration of the waste program. Lastly, in some cases states may be unable to carry out a portion of an RCRA program, and can voluntarily transfer the program back to EPA.

ENFORCEMENT

The enforcement scheme under RCRA consists of several levels of tribunals and penalties, depending on the severity of the violation. An enforcement action may be triggered by an EPA investigation, which is a mandatory duty for the agency under the 1984 amendments. In addition, citizens are provided the power to sue for violations of RCRA orders or if they identify conditions at a site that pose an imminent hazard.

To start the process, RCRA requires the agency (or a state authority) to conduct periodic inspections of waste facilities (42 U.S.C. 6927). These inspections consist of reviewing facility records and entering the facility to take samples and observe operations. Inspections of private waste facilities should take place every two years, according to the statute.

If the inspection identifies problems at the facility, the agency may then proceed with an enforcement action. Such actions range from a simple letter or call from EPA notifying the violator to a criminal action in court leading to fines and jail time (42 US 6928).

The lowest level of enforcement, an administrative action, can assume several forms. An informal administrative action, such as a letter or phone call, would be used to notify a facility that may have committed a relatively minor violation. If that facility fails to come into compliance within the period required by EPA, stronger action may follow.

At the next level, EPA may issue a compliance order requiring the facility to immediately comply with RCRA or to follow a timetable laid out in the order. Terms of the order can include a penalty of up to $25,000 for each day of violation and suspension or revocation of the facility's permit. Those who receive a compliance order can ask the agency for a formal administrative hearing on any allegations in the order.

RCRA also allows the agency to issue an order requiring corrective action at a facility if there has been a release or threat of release from it. In addition, EPA can extend the reach of corrective action to include areas beyond the boundary of the facility (42 U.S.C. 6924(v)).

If a violation is particularly serious or the violator has failed to comply with an administrative order, EPA may choose to pursue a civil enforcement action in court. Civil suits include (1) compliance actions, (2) corrective actions, (3) monitoring and analysis, and (4) imminent hazard actions. Maximum penalties can range from $5,000 per day of noncompliance with the last two categories, while penalties of up to $25,000 per day per violation can be ordered for failure to comply with RCRA regulations. With compliance and corrective actions, EPA may force a violator to take measures to correct shortcomings in waste handling or reduce environmental threats posed by an operation.

Civil suits will typically consist of a combination of actions. For example, in March 1989 a federal district court closed down a landfill in Indiana that had failed to meet RCRA requirements by installing monitoring wells in the wrong locations, failing to cover waste areas to prevent contaminated dust from blowing off, and failing to show adequate financial resources (*U.S. v. Environmental Waste Control Inc.*, 710 F. Supp. 1172, 29 ER Cases 1757, (N.D. Ind. 1989)). The district court imposed a $2.78 million fine and ordered the facility owners and operators to carry out a corrective action plan designed by EPA. This case is a good example of what occurs when someone flagrantly violates RCRA and poses a health hazard in doing so. More commonly, a civil action may be settled through a consent decree signed by EPA and the alleged violator, with penalties and corrective measures designated in a contract subject to a court's approval.

For extremely serious violations, the federal or state government can prosecute under the criminal provisions of RCRA. While used only rarely, the penalties are high: $50,000 per day or a prison sentence of up to 5 years and a total fine of $1 million. Criminal sanctions are invoked in circumstances where there is a "knowing" violation such as handling of hazardous waste without a permit, transport of waste to a nonpermitted facility, or omission of important information in a waste manifest. If the violations place a person in imminent danger, the penalties will reflect the hazard.

Besides federal and state authorities, RCRA allows citizens to sue to enforce the statute (42 U.S.C. 6972). Such suits can involve violations of RCRA regulations, an order issued to a facility, or instances where facility conditions pose an imminent danger. In most cases, the citizen or group must first give notice to the alleged violator, EPA, and the appropriate state agency up to 90 days before proceeding with the suit. In addition, the suit can be precluded if the federal or state government proceeds with an enforcement action during the notice period.

II. CERCLA

INTRODUCTION

The federal environmental law that has caused the most fear in industry, for good reason, is the Comprehensive Environmental Response, Compensation, and Liability Act (CERCLA, or the Superfund law). It imposes severe liabilities on companies and individuals who generate hazardous waste, transport it to a hazardous waste site, or operate the site itself. Other parties also get caught in the liability scheme, such as state and local governments that own or operate a waste site and banks that loan money for a hazardous waste operation.

Congress, when writing the law in 1980, structured it with two guiding principles. First, CERCLA was designed as a spur to site cleanup by ordering the U.S. government to identify where the waste site problem existed and providing the government with the money and regulatory tools to tackle cleanup of those problem areas. Second, Congress wanted to ensure that the site polluters paid for the hazardous waste problem, without leaving the federal government and taxpayers with the bill for the cleanup program.

Under the law:

- The Environmental Protection Agency (EPA), with state help, can select and rank hazardous waste sites across the United States that are in need of attention, placing the most hazardous sites on the National Priorities List (NPL).

- These sites are cleaned up according to the general principles set out by EPA in the National Contingency Plan (NCP), which provides guidelines for remedial investigations and feasibility studies to be done prior to cleanup. The NCP also provides standards for determining the appropriate level of waste cleanup and specifying the extent to which states can participate in cleanup decisions.

- EPA can use money in the trust fund established under CERCLA—the Superfund—to clean up the sites. Then, along with the Justice Department, the government can use enforcement authorities granted in the law to sue parties responsible for site conditions to recover site response costs.

- The agency can also issue administrative orders to responsible parties, requiring them to carry out the site cleanup from the beginning and threatening severe penalties if they ignore the order.

- States are given authority to help EPA determine the appropriate remedy for a waste site, participate in cleanup funding, and sue to recover their cleanup costs.

The Superfund law was one of the most politically charged acts of the congressional session that ended in 1980. Designed to provide a remedy for extensive hazardous waste disposal problems, the CERCLA program quickly became bogged down in an atmosphere of bureaucratic and legal complications.

At the law's 1980 enactment, Congress set aside a $1.6 billion trust fund (the Superfund) to finance cleanup of waste sites. As with any other federal program where large amounts of money are at stake, there was substantial potential for abuse. The history of this fund includes charges of political favoritism in the way the money was allocated and in the way the federal government chose to recover the money from the private sector. The most publicized example involved Rita Lavelle, the first chief of the Superfund program. She was eventually jailed in connection with misuse of her ties to a former employer that was also a liable party at a major hazardous waste site in California. The repercussions from this incident led to the resignation of EPA Administrator Anne Gorsuch.

Another aspect of the financial problems was that Congress failed to specify what standards of liability should apply to private parties who may have been responsible for waste site conditions. As a result, the government spent years in litigation over liability standards, leading to legal expenditures estimated in the millions of dollars. The standards applying to Superfund cleanups were also the source of substantial dispute. EPA was late in issuing the NCP, and parties facing responsibility for cleanup costs constantly challenged the government's right to use the standards in the NCP.

EPA and the federal government have finally reached a point where many of the contentious liability and enforcement issues under CERCLA have been settled. The picture was also clarified when, in 1986, Congress passed the Superfund Amendments and Reauthorization Act, which provided EPA with $8.5 billion in additional Superfund money. CERCLA is due for reauthorization at the end of September 1994.

SITE SELECTION

What Is a Hazardous Waste Site?

EPA, along with other public and private entities, has developed inventories of hazardous waste sites in the United States, with estimates ranging into the tens of thousands as to the number of potential sites. Before EPA prepared the NPL of the worst hazardous waste sites in the country, the agency sought advice from state authorities on candidate sites—typically consisting of abandoned industrial sites where wastes have been left in leaking drums or in lagoons.

CERCLA requires the agency to use an objective measure in ranking the danger posed by various waste sites. This measure, called the Hazard Ranking System (HRS), is a mathematically based model for assessing sites. The two key factors, the hazardousness of the substances at the site and the potential for release of those substances, form the basis for later cleanup and enforcement action by the federal government.

Under the HRS, the agency reviews a number of factors to derive a numerical score for the hazard posed by a waste site. EPA will consider the potential for release to surface water or groundwater, the toxicity and volume of the hazardous substances, and the potential pathways of human exposure. The score assigned under the model provides EPA with a rough method of measuring the relative dangers posed by different sites. To qualify for inclusion on the NPL, a candidate site must receive an HRS score of 28.5 or greater under the agency's system of measurement. Under the Superfund program, EPA has identified more than 1,200 sites that are particularly in need of cleanup (and thus are on the NPL).

EPA has indicated that the time between discovery of a waste site and later investigation and ranking through the HRS can be as little as two years for sites that are perceived to be especially dangerous. However, sites of lesser priority can take up to four or five years to go through the ranking process.

Hazard Identification

One of the major factors in including a site on the NPL—the inherent hazardousness of the substance—is based on definitions found in other environmental laws. Under CERCLA, hazardous substances are defined by reference to other laws: Section 3001 of the Resource Conservation and Recovery Act (RCRA) (see earlier discussion of RCRA), Sections 307(a) and 311(b)(2)(a) of the Clean Water Act (see Chapter 3), and Section 7 of the Toxic Substances Control Act (TSCA) (see Chapter 1). These substances are typically the remnants of manufacturing processes (e.g., the metal solutions remaining from an electroplating operation or the wastewaters from a chemical production facility). In addition, some substances are designated as hazardous under Section 102 of CERCLA.

Several substances are exempt from being designated as hazardous under CERCLA, and thus are exempt from regulation under the law. The broadest exclusion applies to petroleum and its fractions, including unleaded gasoline (CERCLA Section 101(14)). However, additives to gasoline such as benzene can bring petroleum substances back into the CERCLA regulatory scheme, since these additives are considered hazardous.

Another factor in identifying hazards is determining whether there is a release, or the potential for release, of substances from the site. CERCLA defines a release as "any spilling, leaking, pumping, pouring, emitting, emptying, discharging, injecting, escaping, leaching, dumping, or disposing into the environment (including the abandonment or discarding of barrels, containers, and other closed receptacles containing any hazardous substance or pollutant or contaminant)." Courts have interpreted the requirement liberally, usually because they are faced with factual conditions where evidence of soil or water contamination is clear and triggers a federal or state action at the site.

The same questions that arise in the context of placing a site on the NPL also arise when the federal government or a private party tries to recover the costs of responding to the site from the person or company responsible for site conditions. Courts have said that a plaintiff trying to recover cleanup costs must, at a minimum, allege that those costs were linked to the release or threatened release of hazardous waste (see *Bryant v. Colonial Pipeline Co.,* 699 F. Supp. 546, 27 ER Cases 1982 (W.D. Va., 1987)).

THE CLEANUP PROCESS

Background

Once a site is proposed for placement on the NPL, EPA can begin using money from the Superfund trust fund to respond to the site. In addition, placement on the NPL increases the likelihood that potentially responsible parties will be asked to pay site investigation and cleanup costs.

Each cleanup will be guided by the principles laid out in the NCP, which is EPA's framework for site actions. The NCP was first developed for oil and hazardous substance spills under the Clean Water Act, but became much more comprehensive and complicated under the new scheme for hazardous waste cleanup mandated in CERCLA. The NCP establishes the rules for carrying out site studies, the extent of federal and state government authority to take action, and the specific factors that are involved in designing an appropriate site cleanup plan. For private-party cleanups, compliance with the NCP is used as the basis for determining whether cleanup costs may be recovered.

A preliminary observation should be made: once a site is identified by federal and state authorities as posing a problem, the governments will then turn to private parties and order them to take cleanup action or pay for government expenditures at the site. These parties can include waste generators, transporters, and persons otherwise involved in waste handling. Among the most likely candidates for notification are the present owners and operators of the site.

Although liability issues usually arise at a later point in the cleanup process, it is essential for companies or persons that may be linked to a hazardous waste site to take stock of their connection with the site and consider the responsibilities that may arise. Once notified by a state or federal agency, it generally is in the interest of the potentially responsible party (PRP) to play an active part in the process of remedy selection—for reasons of both economic protection and knowledge about site conditions.

Site Studies

To determine the severity of site problems and begin evaluating cleanup alternatives, the agency will perform a remedial investigation/feasibility study (RI/FS). The RI/FS is a two-part site study, with the first part designed to provide the agency with data about the site so it can determine the extent and pathways of contamination. The second part, the feasibility study, establishes alternatives for potential remedies.

Factors evaluated in the remedial investigation include (40 CFR 300.68):

- Risks posed to local population and the environment, and the potential routes of exposure
- Amount, concentration, and hazardousness of substances found at the site
- Groundwater exposure and uses of the water supply
- Sources of environmental contamination
- Adequacy of existing barriers to prevent migration of hazardous substances, along with a description of present and potential future migration of substances
- Federal or state pollution control standards that may apply to the site, and the extent to which they are being exceeded
- Ability of PRPs to respond to releases by implementing and maintaining a selected remedy
- Use of authorities other than the Superfund to respond to the release

As is clear from the above list, scoping of potential response action is essential to the site investigation. At the next stage of the site review process, the feasibility study, discussion of cleanup alternatives becomes more comprehensive.

In general, cleanups must meet all "applicable or relevant and appropriate" requirements (ARARs) of environmental laws, including standards under any of the other laws discussed in this primer, along with state environmental standards. While the NCP is extremely detailed and complicated in setting out standards, the plan developed to specify the cleanup measures for a particular site will be even more exacting.

One of the main purposes of the feasibility study is to develop a range of cleanup options. The NCP sets out categories of alternatives to be covered in the study, including alternatives for (40 CFR 300.68(f)):

- Treatment and disposal of hazardous waste at an off-site facility
- Attaining or exceeding relevant federal health and environmental requirements
- Protection of public health and reduction of threats from hazardous substances, even where federal standards cannot be met
- No action

Costs must be addressed for each alternative, with the goal of ascertaining which is the most cost-effective and will meet environmental standards. This review must include an analysis of costs for operating and maintaining the potential remedy. Any alternative that far exceeds the cost of others, yet does not provide substantially greater protection of public health and the environment, will be rejected.

In addition, each alternative must be evaluated in terms of the engineering practices used to carry out the remedy. EPA has indicated that innovative or advanced technologies may be introduced in the remedy, but there will be an "emphasis on use of established technology."

The technology applied in site remedies is different from site to site; in fact, the differences in technology and results have been a subject of concern to EPA, Congress, and industry. One of the charges leveled against the agency is that a lack of uniformity in waste site remedies leads to widely divergent standards for different areas. Although a certain degree of flexibility is built into CERCLA, differences in levels of public health and environmental protection from site to site are hard for EPA to explain. In addition, such variations raise grounds for dissent by the industry parties who are likely to be sued by the government for the costs of cleaning up the sites. These PRPs are likely to point out site plan problems to a court when the federal government sues to recover response costs.

Remedies: Emergency, Long-Term

The NCP provides for two levels of cleanup actions: (1) short-term emergency removal actions and (2) long-term remedial actions.

Short-term measures are limited to emergency situations, allowing the agency to take action for a period of one year with a maximum expenditure of $2 million; however, the agency can seek additional time and money if it believes cleanup activity will contribute to a more permanent remedy for the site. These short-term responses are designed to stabilize any emergency situation that arises at a hazardous waste site.

For example, massive rains in an area might lead to flooding, and also cause a lagoon at a waste facility to overflow; the NCP provides that EPA can quickly build a dike to prevent overflow. As a result, the short-term threat is abated, but little may be accomplished to eliminate or substantially reduce risks posed by the site. Other typical emergency activities could include installing security systems around the site or covering dry waste piles to prevent hazardous waste from blowing off. In addition, EPA can respond to emergency situations where single barrels have begun to leak and are contaminating the environment.

Arrangements for long-term cleanup involve a lengthier and more complicated process, with the goal of eliminating the risks posed by hazardous waste. The cleanup standard imposed by the NCP is selection of a "cost-effective remedial measure that effectively mitigates and minimizes threats to and provides adequate protection of public health and welfare and the environment" (40 CFR 300.68(i)(1)).

The statute indicates two other measures for selection of waste sites remedies. Treatment approaches that permanently reduce the volume and mobility of hazardous substances are preferred. In addition, on-site treatment is preferred to off-site transport and disposal of contaminated materials (CERCLA Section 121(a)).

The NCP guidance on site remedies covers air pollution control, surface water controls, groundwater controls, and contaminated water and sewer lines (40 CFR 300.70(b)(1)). It also describes, in a little more detail, the types of remedies that will be acceptable. For example, groundwater controls might include the use of impermeable barriers such as slurry walls or grout curtains, or a groundwater pumping system might be employed. Suggested methods for waste treatment might include some of the specified biological or chemical methods listed in the NCP.

As implied in the standards for scoping alternative remedies, the remedy selected will have to meet the ARARs pertinent to the specific site. However, there are situations where a selected remedy will not have to meet ARARs. This can occur when (40 CFR 300.68(i)(5)):

- An alternative is only a portion of a more comprehensive remedy
- There would be excessive draining of funds from the Superfund trust fund (this exception does not apply where PRPs are paying for the remedy)
- The alternative involves a remedy that is technically impracticable
- Unacceptable adverse environmental impacts might result from carrying out the alternative
- There is a strong public interest in an expedited cleanup and litigation would be appropriate

Once the RI/FS and cleanup alternatives discussed in the study have been submitted for public review, EPA will select a remedy for the specific site. The selected remedy will be described in a document called a record of decision (ROD) that will contain the exact steps for the cleanup and detail the responsibilities of the various government and private parties. Once the ROD is completed, the plan is carried out through the Remedial Design/Remedial Action (RD/RA). This is the stage where the details of each stage of a cleanup are presented and then carried out at the site.

Documentation of costs and site conditions is also important during each step of the remedial process, so that the party paying for the cleanup can back up its claims in a later suit to recover costs under Section 107 of CERCLA. Under the NCP, the documentation must address the source and circumstances of the hazardous conditions at the site, the identity of potentially responsible parties, and an accurate accounting of any costs incurred (40 CFR 300.69(a)).

With the completion of site cleanup, EPA requires that a site continue to be monitored for at least 30 years to ensure that the remedy has been successful. A site review will take place every five years as part of Operation and Maintenance after the cleanup, and changes can be required as a result of these reviews. In addition, the parties will be responsible for monitoring site conditions and tackling maintenance of cleanup equipment and site status.

PRP Involvement in Remedy Selection

Once a PRP receives a notice from EPA saying that it has contributed to conditions at a site where the agency is contemplating a Superfund cleanup, the party should do its best to obtain information about EPA's remedy selection process. Notice will usually involve an administrative enforcement order issued under Section 106 of CERCLA, or may be inferred through EPA requests for information about a PRP's activities at a facility. At this point, the PRP should be aggressive in obtaining copies of any public reports and, where possible, documents collected by EPA to track the amounts and types of wastes and the parties believed to be responsible for site conditions. Once the cleanup process begins, a PRP is well advised to keep track of that process and submit comments on the administrative record developed by the agency. This protects a PRP in case it wants to raise challenges to the cleanup plan in a later suit.

PRPs are allowed to conduct the site studies, but must adhere to EPA requirements. Plans for the RI/FS must first be submitted to the agency for approval, with the expectation that state or federal officials will want to establish some form of oversight to monitor the progress and accuracy of the studies. EPA has been somewhat reticent about allowing PRPs to take on the RI/FS process; to date, EPA-led site studies have prevailed over PRP-led studies by a 5:1 ratio.

State Responsibilities

Unlike other environmental statutes, states' responsibilities under CERCLA are rather limited. The CERCLA program remains primarily federal in nature, and states are not delegated authority to operate to the same extent as in, for example, the Clean Water Act or the Clean Air Act.

However, states play a significant role in helping EPA identify priority waste sites and in selecting an appropriate remedy. Under Section 121 of CERCLA, state standards may be applied in designing the remedy for a site. If the state standard is stricter than the federal requirements that would otherwise apply, the state must make up the difference in cost for the remedy. States are also limited in their authority to impose standards for on-site cleanup activities—EPA is not required to obtain state permits for activity taking place on-site.

States have extensive opportunities to participate in remedy selection by commenting on EPA proposals and participating in negotiations with PRPs. States are also expected to provide at least 10 percent of the funding for site response when EPA initiates a cleanup using Superfund money. Most states have passed their own "mini-superfund" statutes to collect funds for state efforts to remedy hazardous waste problems.

Court Review of Remedies

Once EPA selects a remedy for a waste site, courts are unwilling to review the selection process. A string of court decisions has produced the general rule that courts will refuse to review a cleanup remedy until EPA takes steps to order a party to carry out the remedy or tries to recover its own response costs.

This lack of "preenforcement review" is based on policy considerations. As one court indicated, "Congress envisioned a procedure that permits the EPA to move expeditiously in the face of a potential environmental disaster." Court delay could lead to further harm from waste site conditions, Congress warned (*Wagner Seed Co. v. EPA,* 800 F.2d 310, 24 ER Cases 1916 (2d Cir., 1986)).

Once the federal government decides to proceed with an enforcement action, PRPs are entitled to limited court review of the remedy. In the 1986 amendments to CERCLA, Congress limited the scope of judicial review to information found in the administrative record developed by EPA in the ROD for a site (CERCLA Section 113).

ENFORCEMENT

Background

Once EPA has gone through the process of selecting sites for cleanup, designing a remedy for hazardous waste leakage, and beginning to implement the remedy, the next question is who will actually pay for the cleanup. Even though Congress set aside an $8.5 billion fund for cleanups in the 1986 CERCLA amendments, the emphasis was on making sure that the federal government forces the polluter to pay. EPA has the option to undertake cleanup using Superfund money but has made it clear that it would prefer using the funds of private parties to tackle waste sites. Congress has provided the agency with substantial authority to fulfill this goal.

There are two major enforcement mechanisms available to EPA under CERCLA: (1) issuing administrative orders to PRPs under Section 106 ordering them to undertake a site cleanup, and (2) filing suit under CERCLA Section 107 to obtain a court-ordered cleanup or recover response costs already spent at a site.

Section 106: EPA Administrative Orders

Under Section 106, the agency can unilaterally issue orders to parties it believes are responsible for conditions at a hazardous waste site. With these orders, EPA can require PRPs to abate conditions believed to pose a danger to public health and the environment. This authority is used most frequently when EPA begins to undertake site studies and remedial activity, and it encourages PRPs to become involved in the process early. As the agency gathers information about site conditions, it will exercise its option to bring more PRPs into the process through Section 106 orders.

The early years of the Superfund program saw a tendency for PRPs to be reticent about responding to these administrative orders; they preferred instead to wait for EPA to spend its own funds, and then have to sue PRPs later and prove liability under Section 107. With the evolution of stringent liability standards and the agency's increasing sophistication in proving the responsibility of individual PRPs, administrative orders have served to spur more cleanup activity by PRPs.

To back up the Section 106 authority, CERCLA authorizes the agency to assess civil penalties against persons or companies who fail to comply with the order "without sufficient cause." This means that a party receiving an EPA order must determine whether it can avail itself of the defenses allowed if the U.S. government were to sue it later for response costs under Section 107 of CERCLA. The PRP would have to make a "good-faith" claim that site conditions were caused by (1) an act of God, (2) an act of war, or (3) the acts or omissions of third parties at the site. If the party is clearly responsible for site conditions and knows that the defenses do not apply, it faces a great risk by violating the administrative order.

Penalties for such violations can range up to $25,000 per day, to be imposed in a judicial enforcement action. Failure to comply with a Section 106 order can also lead to punitive damages being assessed against the noncomplying party. These penalties can be triple the amount the U.S. government spent in response costs at a site. It should be noted that treble damages are not automatically assessed; in fact, the agency must sue to obtain such penalties under Section 107 of CERCLA. EPA has done so in less than 20 cases, and, as of November 1989, had succeeded in obtaining punitive damages in only one case.

The dilemma for a party at a Superfund site is that it can either choose to undertake a cleanup that may cost millions of dollars or risk substantial penalties for failing to comply with the Section 106 order. One court has described the situation as one in which the PRP "thinks itself caught between the devil and the deep blue sea" (see *Wagner Seed Co. v. EPA,* 800 F.2d 310, 24 ER Cases 1916 (2d Cir., 1986)).

Section 107: Going to Court

What happens when EPA or a state goes forward with a cleanup and then wants to recover its money from the parties responsible for site conditions? The answer is contained in CERCLA Section 107. The federal and state governments will sue PRPs linked to the site, framing the suit to match the elements of liability identified in the Section 107 scheme. The decision as to which PRP to pursue is left to the discretion of the governments; however, the U.S. government has had few qualms about suing virtually all of the PRPs at a site, as shown by the hundreds of parties that may be defendants in a single Superfund suit.

Similarly, private companies or individuals can also use authorities under Section 107 to recover money spent in a site cleanup. If a PRP ends up reimbursing the federal or state government for a cleanup, the PRP can still go after other parties at the site to recover their share of cleanup costs. In this instance, cost sharing is called "contribution" among responsible parties at a site, and is dealt with in a later section of this chapter.

Elements of a Section 107 Enforcement Suit

In a federal or state government cost-recovery suit, the following must be proved:

- That a PRP fits within a class of liable parties identified in Section 107. These include (1) the current owner or operator of a facility holding hazardous substances, (2) previous owners or operators who were assoicated with a facility when hazardous substances were disposed of there, (3) generators of hazardous substances, and (4) transporters of hazardous substances.
- That there has been a release or a threat of release of hazardous substances from the facility.
- That costs have been incurred in responding to the release.
- That these costs are eligible for payment under CERCLA.

Liable Parties

The universe of potential parties in a CERCLA suit is large; all of the types of parties identified above have been held liable for cleanup costs under Section 107. The central question is the nexus between the party and the contaminated property. When ordering a site cleanup, EPA will frequently have hundreds of parties to turn to in sending Section 106 administrative orders, and these same parties may be the ones EPA sues in a later action under Section 107. At the same time, a Superfund cleanup may be triggered by improper placement of waste by a single manufacturer on a site.

Courts have been willing to extend the concept of ownership beyond the obvious bounds, going so far as to impose liability on banks that foreclose on a mortgage for property containing a waste site. In other cases, liability has been imposed on companies that acquire subsidiary firms with contaminated property.

Furthermore, CERCLA liability covers both past and present owners and operators of a waste facility. One of the few limits on CERCLA liability is that prior owners and operators will be held liable only if they were connected with the facility at the time a hazardous substance was disposed of there. This limitation does not apply to current owners or operators, who can be held liable without proof that they caused waste problems (see *New York v. Shore Realty Corp.*, 759 F.2d 1032, 22 ER Cases 1625 (2d. Cir., 1985)).

Generators of waste also are subject to liability based on a minimal showing of causation. One of the most common defenses raised by generators is that the government, in seeking response costs, cannot show that the generator's wastes actually caused the problem at the waste site. Federal courts have uniformly rejected this approach and allowed the government to show simply that substances such as those disposed of by the generator contributed to site conditions (see *U.S. v. Monsanto Co.*, 85 F.2d 160, 28 ER Cases 1177 (4th Cir. 1988)).

Release or Threatened Release

As explained early in the chapter, EPA looks at the potential for releases of hazardous waste from a site when deciding whether the site merits attention under CERCLA. The same factors apply in establishing that there has been a release or threat of release from the site, including how hazardous the substance is; whether it has contaminated soil, surface water, or drinking water supplies; how securely the hazardous substance was placed at the site; whether the method of storage was designed to limit contamination; and whether means exist to determine how far contamination has spread.

At a minimum, the plaintiff suing to recover cleanup costs must allege that those costs were linked to the release or threatened release of hazardous waste (see *Bryant v. Co-*

lonial Pipeline, 699 F. Supp. 546, 27 ER Cases 1982 (W.D. Va. 1987)).

Response Costs

An array of costs can be recovered by government or private parties that respond to conditions at a hazardous waste site. All costs of a removal or remedial action may be recovered under CERCLA Section 107(a) by a federal or state government, as long as they are consistent with the NCP. In addition, private parties can recover costs consistent with the NCP, as long as they are "necessary costs of response" that can cover the same types of remedial actions as those of a federal or state government. Other response costs may also be sought, including the costs of responding to natural resource damages and any costs of health assessments or health effects studies carried out under CERCLA. Courts have also been amenable to suits where government and private plaintiffs have attempted to recover the costs of investigating site conditions. Although there have been few cases in which courts have rejected the costs claimed in a Superfund suit, there have been mixed rulings on the ability of residents living near a Superfund site to recover the costs of medical monitoring that might not be part of the regular CERCLA process for assessing the health effects of a site.

It should be noted that in cases where private parties are seeking to recover response costs, such parties are not required to obtain the prior approval of the federal or state government to go forward with a suit. Courts have adopted this approach based on EPA's own interpretation of CERCLA, as stated in the NCP (see *Wickland Oil Terminal v. Asarco Inc.*, 792 F.2d 887, 24 ER Cases 1545 (9th Cir., 1986)).

Standard of Liability

The most contentious aspects of CERCLA are the liability provisions. The original 1980 law was poorly written, which left federal courts with the task of trying to determine how Congress intended to define who could be held liable for cleanup costs. In the end, a very broad set of standards were adopted—strict, joint and several liability—which have caused industry much misery and expense.

Under strict, joint and several liability, one company can be held liable for the entire cost of a site cleanup, no matter how many other parties may also be responsible for the site and no matter how little waste the liable company contributed to the site. "Strict" liability refers to the fact that only a small amount of evidence is needed to establish that a party was linked to the site. "Joint and several" liability applies where a number of parties may be responsible for the site, yet the conditions at the site make it so difficult to determine who contributed what substance that a court is willing to hold one party responsible for the entire amount of cleanup costs.

With broad liability to spur action by parties responsible for site conditions, the federal government has spent funds from the Superfund, with the expectation that the money can be recovered fairly quickly, under Section 107 of CERCLA, from any one of the parties responsible for site conditions. However, the last 10 years have shown that any company or individual who has been the subject of a suit for response costs has fought aggressively to prevent the imposition of liability.

Going After Other PRPs

Once strict, joint and several liability is imposed against one or more parties at a hazardous waste site, Section 133(f) of CERCLA provides that the liable parties may sue other parties for contribution of response costs.

In reviewing claims for contribution against PRPs, courts are to "allocate response costs among liable parties using such equitable factors as the court determines are appropriate." These factors may include the relative fault of the parties for site conditions, the volume of waste deposited, the relative toxicity of waste, and the existence of any agreements between the parties that might act to limit liability. In general, though, courts are disinclined to become entangled in the scientific and complicated factual determinations that arise in apportioning liability, and they have indicated a preference for settlements among PRPs.

Courts have responded to PRPs' arguments about "innocence" or fault for site conditions by determining that such factors do not prevent liability from attaching. However, the factors that would apply to a claim of innocence (e.g., amounts and toxicity of waste disposed of) are the same as those considered by a court when addressing the question of how to apportion costs (see *Chemical Waste Management Inc. v. Armstrong World Industries Inc.*, 699 F. Supp. 1285, 27 ER Cases 1097 (E.D. Pa. 1987)).

Defenses to Liability

In the wake of the stringent liability standards imposed by CERCLA and the courts, very few defenses to liability have been allowed. As noted earlier, the original statute established only three acceptable defenses (CERCLA Section 107(b)): (1) act of god, (2) act of war, and (3) acts or omissions by third parties. Later, the 1986 amendments somewhat expanded the reach of the third-party defense. Congress set up a framework of conditions under which an "innocent landowner" would be protected from CERCLA liability.

The defense of war conditions has obvious limitations, and is rarely raised in CERCLA suits. Blaming site conditions on acts of God has been attempted in several cases, but courts have chosen to take a narrow view of such attempts. For example, a PRP argued in one case that heavy rainfall contributed to problems at a waste site in California. The federal court determined that only "exceptional natural phenomena" justify the defense: "the rains were foreseeable based on normal climatic conditions and any harm caused by the rain could have been prevented." In addition, according to the court, natural acts were not the sole reason the site had problems (*U.S. v. Stringfellow,* 661 F. Supp. 1053, 26 ER Cases 1624 (D.D.C. Cal. 1987)).

The Most-Used Defense: Third-Party Defense

The third-party defense is also limited in application. To avail itself of this defense, a PRP must show that there was no contract between it and the third party that led to the waste site conditions, either directly or indirectly. In addition, the PRP has to show that "he exercised due care with respect to the hazardous substance concerned" and took precautions against the foreseeable actions or omissions of third parties (CERCLA Section 107(b)(3)).

Again, courts have examined the question of whether hazardous waste activities could have been foreseen in the relationship between the PRP and the third party. In one instance, a court chided a company for failing to take due care in complying with state orders to respond to problems with a waste site and for failing to foresee that a related firm was dumping wastes at the site (*U.S. v. Tyson,* 25 ER Cases 1897 (E.D. Pa. 1986)).

The third-party defense was extended in the 1986 amendments to address the perception that landowners were unfairly being held accountable for the practices of past site owners. Congress addressed this inequity by creating the "innocent landowner" defense in CERCLA Section 101(35)(A). A PRP will be able to use the defense if it acquired property after hazardous substances were disposed of, and the PRP can show that at the time it acquired the property it "did not know and had no reason to know that any hazardous substance which is the subject of the release or threatened release was disposed of on, in, or at the facility."

What makes the burden particularly hard on a PRP is that the statute specifies that at the time the property was acquired, the PRP must have undertaken "all appropriate inquiry into the previous ownership and uses of the property consistent with good commercial or customary practice." CERCLA directs a court considering the defense to look at "any specialized knowledge or experience on the part of the defendant, the relationship of the purchase price to the value of the property if uncontaminated," and other factors concerning a purchaser's potential knowledge of site contamination.

As a practical matter, this new provision requires a prospective purchaser to undertake some form of an environmental inquiry, such as an environmental assessment. Although the process is time consuming and sometimes costly, it at least helps provide the purchaser with evidence to support use of the "innocent landowner" defense if problems later appear.

An environmental assessment typically begins with three sets of actions: (1) review of government and public records pertaining to environmental compliance and uses of the property; (2) review of site history through information held by the site owner, state or local government, and employees of the owner; and (3) site walk-through and inspection of property for potential contamination. If there are signs that wastes were handled at the site, further testing and monitoring of soil and of surface water and groundwater may be necessary.

On the other hand, at least one PRP has succeeded in raising the argument that failure to inspect newly purchased property was within commercial practices. Although the court did not conclude that the failure to inspect was appropriate, it opted to allow further development of the factual record in the case instead of automatically rejecting the claim (*U.S. v. Serafini,* 706 F. Supp. 346, 28 ER Cases 1162 (M.D. Pa. 1988)).

Settlement Authorities

Under the 1986 amendments, settlements are encouraged between the federal and state governments and potentially responsible parties as an alternative to the financial drain caused by litigation. In reality, PRPs have been so wary of each other's and the government's motives as to make site settlements less likely as the number of PRPs involved in the case increases. As mentioned earlier, it is not unusual to have several hundred parties associated with one hazardous waste site.

Although there have been recent signs of progress in negotiations over liability, with a larger number of settlements reported by the federal government, the progress of site cleanups has been slow. Of the almost 1200 sites slated for priority action since 1980 only 40 sites have been removed from the NPL. However, an EPA official predicted that cleanups would be completed at 130 NPL sites by the end of fiscal year 1992, along with 200 sites to be finished by the end of fiscal year 1993. However, in 1989 there were 218 site settlements involving the federal government, worth more than $1 billion, according to a top EPA enforcement official.

Section 122 of CERCLA states the conditions and authorities available in settlements with the federal government. Settlements are to be made at the discretion of the United States, but the statute indicates that they are to be carried out "[w]henever practicable and in the public interest." If the government decides not to settle with PRPs, it must notify the parties and explain why the decision was reached. Dissatisfied PRPs can do little to question that decision; Section 122(a) provides that the government's decision is not subject to judicial review.

If the government decides to proceed with a settlement, and initiates negotiations with PRPs, the statute provides that potential parties to the agreement should be given information. That information includes the names and addresses of PRPs, the volume and types of substances contributed by the various parties, and a ranking by volume of substances at the Superfund site. In addition, once negotiations begin there is a delay imposed on further EPA administrative or judicial enforcement action or studying of the site (Section 122(e)(2)).

Settlement terms can cover factors such as allocation of costs among parties, identification of cleanup responsibilities, and any other aspect of the cleanup and site monitoring process. One of the most important aspects of a settlement is that it can include a promise by the federal government that it will not sue a settling party for any amount above what was already stipulated in the agreement. A "covenant not to sue" is subject to a number of considerations and limitations explained in Section 122(f), including:

- The effectiveness of the site remedy
- The nature of risks posed by hazardous wastes at the site
- Whether a substantial reduction of wastes has been achieved
- The sources of funding that would be used if additional remedial action was needed for the site
- Whether the remedy will be carried out by PRPs

The U.S. government may also reserve the right to reopen the settlement if the costs of responding to site conditions substantially exceed the original expectations of the government.

The settlement can also provide protection against contribution actions by other PRPs that did not participate in the settlement. Nonsettling parties will have the opportunity to comment on agreements that are reached in most actions where the statute requires an opportunity for public comment before the settlement is approved by a federal court (Section 122(d)(2)). However, Congress attempted to make it attractive for PRPs to settle with the federal government.

Two special types of settlement provisions were created by Congress in CERCLA. One provision places a special emphasis on forcing the government to settle with parties that would be responsible for "only a minor portion of response costs"—so-called "de minimis" parties (Section 122(g)). A federal court in New Jersey recently determined that Congress encouraged the use of such settlements as a timely way to respond to waste sites, and accepted a $3 million de minimis settlement at a landfill because:

- EPA's administrative record supported the determination that the 12 settling parties were responsible for about 3 percent of the waste at the site
- The agreement was reached after "arm's-length" negotiations between the parties
- The settlement could be reopened if evidence showed the parties contributed greater amounts to site conditions than originally believed
- The settlement averted lengthy litigation while providing additional funds for the U.S. government to respond to the site. (*U.S. v. Rohm & Haas Co.*, 712 F. Supp. 666, 30 ER Cases 1520 (D.C. N.J. 1989)).

Another settlement provision allows the government to make a "nonbinding preliminary allocation of responsibility" (NBAR) to stimulate settlement by providing an early indication to PRPs about cost allocation (Section 122(e)(3)). According to an EPA enforcement official, this section has been used only sporadically.

III. PESTICIDE DISPOSAL UNDER FIFRA

The Federal Insecticide, Fungicide, and Rodenticide Act (FIFRA) requires that pesticide registrants provide EPA information on safe storage and disposal methods. This information may be required under Section 3 (product registration), Section 6 (registration cancellation and suspension). FIFRA Section 19 includes requirements for pesticide storage, disposal, transportation, recall, and container design.

Most registered pesticides include disposal instructions on their labels. Pesticides classified as hazardous waste must be disposed of according to rules issued under the Resource Conservation and Recovery Act (RCRA). (See the earlier discussion in this chapter of hazardous waste disposal rules.)

Individuals disposing of pesticides not covered by RCRA regulations and labeling instructions must, under 40 CFR 165, follow the use, storage, and disposal instructions on the pesticide label.

Any undiluted spray material that will not be used should be returned to the dealer, who may be able to dispose of the product or ship the product back to the manufacturer.

Containers always should be triple-rinsed and returned to the dealer for reuse or reconditioning. Some landfills accept triple-rinsed pesticide containers.

Pesticides should not be disposed of by open dumping, open burning, or ocean dumping. The preferred method of disposal for organic pesticides is incineration. If a suitable incineration facility is not available, the product should be buried in a designated landfill. If pesticides are disposed of in a landfill, records must be maintained to permit location and retrieval.

Pesticides containing certain heavy metals or metallo-organic products should be converted into nonhazardous compounds and the heavy metals recovered. If conversion and recovery facilities are not available, the pesticides should be encapsulated and buried in a designated landfill.

Disposal of Canceled or Suspended Products

The 1988 amendments to FIFRA eliminated the requirement that EPA store and dispose of suspended and canceled pesticides at taxpayers' expense. EPA was criticized for how it handled the disposal of three banned products: 2,4,5-T/Silvex, ethylene dibromide, and dinoseb. The revised storage and disposal provision requires EPA and the registrant to share responsibility for proper disposal.

EPA may require registrants and distributors to recall canceled and suspended products under FIFRA Section 6. The agency may require companies that sell, distribute, or commercially use pesticides to demonstrate the financial ability to carry out recalls. Recalls may be mandatory or voluntary. Companies may be eligible for some storage cost reimbursement in certain recall situations.

Pesticide Containers

Disposal of empty pesticide containers and design requirements to promote the safe storage and disposal of pesticides are mandated under FIFRA Section 19. In 1988, Congress directed EPA to require recycling, reuse, and disposal of pesticide containers by 1991 with rules that would become effective in 1993.

FIFRA also requires EPA to regulate container storage, transport, and disposal and to develop procedures and standards for the removal of pesticides from containers prior to disposal. These rules are scheduled to take effect by 1993 (40 CFR 165).

IV. RELATIONSHIP BETWEEN SDWA STANDARDS AND DISPOSAL LAWS

Safe Drinking Water Act (SDWA) standards play an important role in waste disposal and cleanup programs because the act was designed to protect groundwater, and groundwater contamination is one of the most common consequences of improper waste disposal.

Section 121 of the Comprehensive Environmental Response, Compensation, and Liability Act, which covers cleanup standards, requires compliance with SDWA (and other pollution laws). It also requires that remedial actions attain SDWA maximum contaminant level goals for each pollutant at the site.

Section 1006 of the Resource Conservation and Recovery Act (RCRA) prohibits states and other authorities from adopting waste programs that are inconsistent with SDWA and other laws. It also requires the EPA administrator to integrate SDWA requirements into RCRA standards.

RCRA Section 3004, which addresses RCRA-permitted facilities, prohibits disposal of wastes that risk contamination of underground drinking water supplies. It also adopts the SDWA definition of "underground source of drinking water." In addition, Sections 264.92–264.94 of the RCRA regulations impose extensive groundwater protection requirements.

APPENDIX A

AIR QUALITY

Air Concentrations

Particulate Matter 288

Sulfur Dioxide 290

Carbon Monoxide 292

Nitrogen Dioxide 294

Ozone 296

Lead 298

Air Concentration Maps

PM-10 300

Sulfur Dioxide 301

Carbon Monoxide 302

Nitrogen Dioxide 303

Ozone 304

Lead Concentration 305

Tables

National Sulfur Oxides Emission Estimates, 1981-1990 306

National Carbon Monoxide Emission Estimates, 1981-1990 306

National Nitrogen Oxides Emission Estimates, 1981-1990 307

National Volatile Organic Compound Emission Estimates, 1981-1990 307

National Lead Emission Estimates, 1981-1990 308

National Total Particulate Emission Estimates, 1981-1990 308

National PM-10 Emission Estimates, 1985-1990 309

PARTICULATE MATTER

AIR CONCENTRATIONS : Total Suspended Particulates (TSP) and PM-10

1982-90*: 3 percent decrease **TSP** (based on geometric mean at 1265 sites)
* 1981 data affected by a change in filters

1989-90: 3 percent decrease **TSP** (based on geometric mean at 734 sites)
8 percent decrease **PM-10** (based on arithmetic mean at 339 sites)

EMISSIONS : Total Particulates (TP) and PM-10

1981-90: 6 percent decrease **(TP)**
(Note: 9-year 1982-90 change was 6 percent increase)

1989-90: 4 percent increase **(TP)**; 5 percent increase **(PM-10)**

OVERVIEW

Trends TP emissions from historically inventoried sources have been reduced 59 percent since 1970. During the 1980's, TSP air quality levels improved 3 percent. In 1987, EPA replaced the earlier TSP standard with a PM-10 standard. (PM-10 focuses on the smaller particles likely to be responsible for adverse health effects because of their ability to reach the lower regions of the respiratory tract.) Ambient monitoring networks have recently been revised to measure PM-10 rather than TSP. Although PM-10 trends data are limited, ambient levels decreased 11 percent between 1988 and 1990. The PM-10 portion of TP emissions is estimated to have increased 7 percent since 1985 due to increases from transportation sources and forest fires. Nationally, fugitive sources provide 6-8 times more tonnage of PM-10 emissions than historically inventoried sources.

Status In October 1991, EPA designated 70 areas as nonattainment for PM-10. National average TSP levels in 1990 were the lowest of the past decade. Comparing 1989 and 1990, most of the country experienced an increase in precipitation and a decrease in TSP and PM-10.

1990 Clean Air Act The Act focuses attention on nonattainment of PM-10 health based standards. The Acid Rain provisions of the Act address visibility impairment caused by fine (<2.5 micrometer) particles.

TSP TREND, 1981-1990
(ANNUAL GEOMETRIC MEAN)

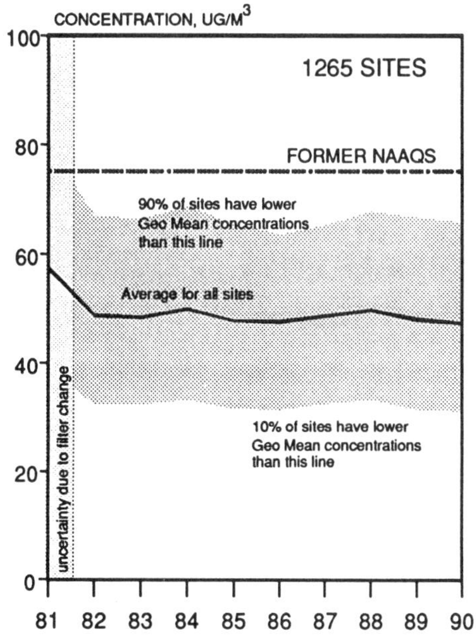

TP EMISSIONS TREND
(1981 vs. 1990)

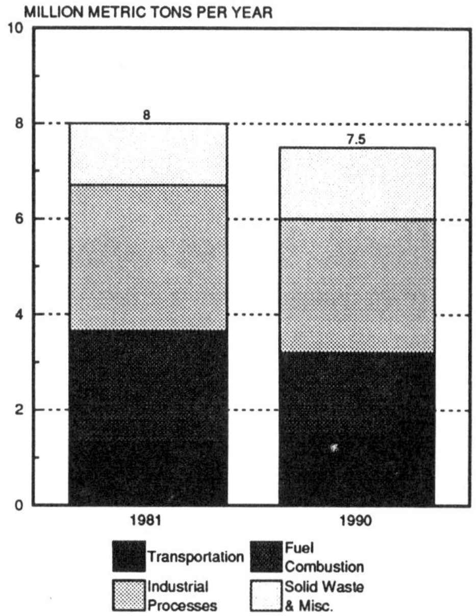

PM EFFECTS

Based on studies of human populations exposed to high concentrations of particles (often in the presence of sulfur dioxide), and laboratory studies of animals and humans, the major effects of concern for human health include effects on breathing and respiratory symptoms, aggravation of existing respiratory and cardiovascular disease, alterations in the body's defense systems against foreign materials, damage to lung tissue, carcinogenesis and premature mortality. The major subgroups of the population that appear likely to be most sensitive to the effects of particulate matter include individuals with chronic obstructive pulmonary or cardiovascular disease, individuals with influenza, asthmatics, the elderly and children. Particulate matter causes damage to materials, soiling and is a major cause of substantial visibility impairment in many parts of the U.S.

Source: U.S. Environmental Protection Agency, *National Air Quality and Emissions Trends Report, 1990*, Nov. 1991, pp. 1-4–1-5.

SULFUR DIOXIDE (SO_2)

AIR CONCENTRATIONS

1981-90: 24 percent decrease (arithmetic mean at 457 sites)
30 percent decrease (24-hour second high at 452 sites)
87 percent decrease (24-hour exceedances at 452 sites)

1989-90: 7 percent decrease (arithmetic mean at 552 sites)

EMISSIONS : SO_x

1981-90: 6 percent decrease

1989-90: 2 percent increase

OVERVIEW

Trends SO_x emissions decreased 25 percent since 1970. During the 1980's, emissions improved 6 percent while average air quality improved by 24 percent. This difference occurs because the historical ambient monitoring networks were population-oriented while the major emission sources tend to be in less populated areas. The exceedance trend is dominated by source oriented sites. The 1981-90 decrease in emissions reflects reductions at coal-fired power plants. The 1989-90 emissions increase is due to increases from fuel combustion.

Status Almost all monitors in U.S. urban areas meet EPA's ambient SO_2 standards. Dispersion models are commonly used to assess ambient SO_2 problems around point sources because it is frequently impractical to operate enough monitors to provide a complete air quality assessment. Currently, there are 50 areas designated nonattainment for SO_2. Current concerns focus on major emitters, total atmospheric loadings and the possible need for a shorter-term (i.e. 1-hour) standard. Seventy percent of all national SO_x emissions are generated by electric utilities (92% of which come from coal fired power plants).

1990 Clean Air Act The Acid Rain provisions include a goal of reducing SO_x emissions by 10 million tons relative to 1980 levels. The focus in this control program is innovative market-based emission allowances which will provide affected sources flexibility in meeting the mandated emission reductions. This is EPA's first large-scale regulatory use of market-based incentives. These reductions will improve visibility in the East by substantially reducing SO_x emissions. These emissions are transformed into fine acid sulfate aerosol, the main cause of regional visibility impairment.

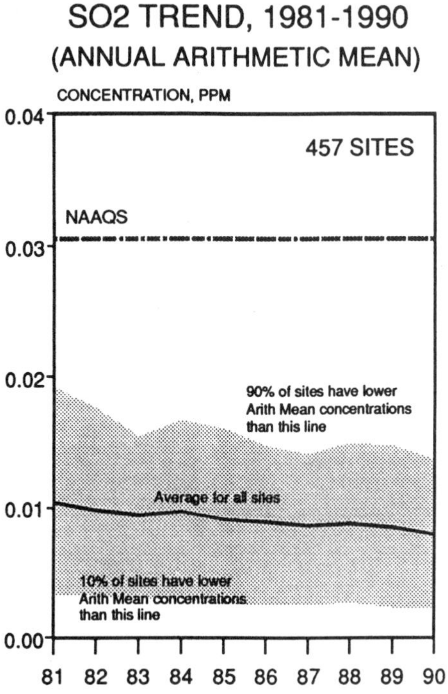

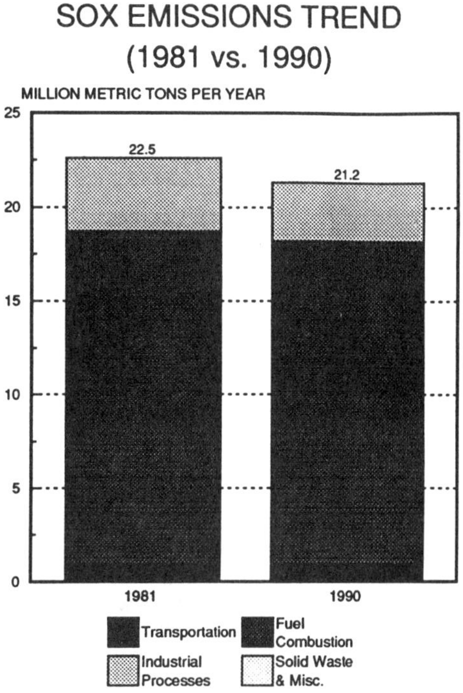

SO_2 EFFECTS

The major health effects of concern associated with high exposures to sulfur dioxide include effects on breathing, respiratory illness and symptoms, alterations in the lung's defenses, aggravation of existing respiratory and cardiovascular disease, and mortality. The major subgroups of the population most sensitive to sulfur dioxide include asthmatics and individuals with chronic lung disease (such as bronchitis or emphysema) or cardiovascular disease. Children and the elderly may also be sensitive. Sulfur dioxide produces foliar damage on trees and agricultural crops. It and nitrogen oxides are major precursors to acidic deposition (acid rain), which is associated with a number of effects including acidification of lakes and streams, accelerated corrosion of buildings and monuments and visibility impairment.

Source: U.S. Environmental Protection Agency, *National Air Quality and Emissions Trends Report, 1990*, Nov. 1991, pp. 1-6–1-7.

CARBON MONOXIDE (CO)

AIR CONCENTRATIONS

1981-90: 29 percent decrease (8-hour second high at 301 sites)
87 percent decrease (8-hour exceedances at 301 sites)

1989-90: 8 percent decrease (8-hour second high at 359 sites)

EMISSIONS

1981-90: 22 percent decrease

1989-90: less than 1 percent decrease

OVERVIEW

Trends Carbon monoxide emissions decreased 41 percent since 1970. Progress continued through the 1980's with 29 percent improvement in air quality levels and a 22 percent reduction in total emissions. This progress occurred despite continued growth in miles of travel in the U.S. Transportation sources account for approximately two-thirds of the nation's CO emissions. Emissions from highway vehicles decreased 37 percent during the 1981-90 period, despite a 37 percent increase in vehicle miles of travel. Estimated nationwide CO emissions decreased less than 1 percent between 1989 and 1990, with forest fire activity in 1990 offsetting the 7 percent decrease in CO emissions from highway vehicles.

Status In October 1991, EPA designated 42 areas as nonattainment for CO.

1990 Clean Air Act The remaining CO nonattainment areas have specific planning and implementation requirements specified in Title I of the Act that vary depending upon the magnitude of the CO problem. In addition, Title II of the Act, which deals with mobile sources, includes a variety of provisions to help reduce CO levels including a winter time oxygenated fuels program for CO nonattainment areas, increased application of vehicle inspection and maintenance programs, and a tailpipe standard for CO under cold temperature conditions.

CO TREND, 1981-1990
(ANNUAL 2ND MAX 8-HR AVG)

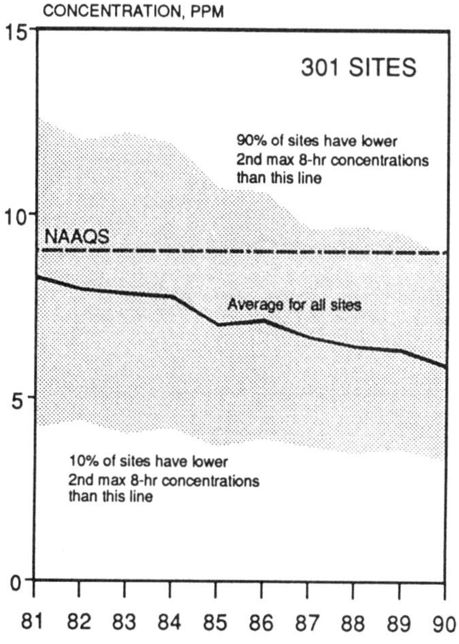

CO EMISSIONS TREND
(1981 vs. 1990)

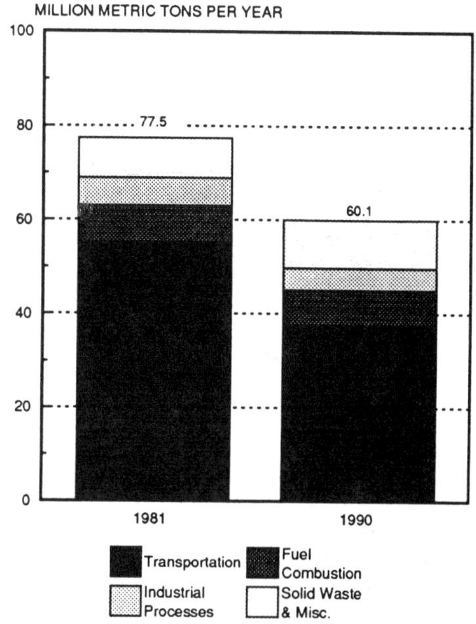

CO EFFECTS

Carbon monoxide enters the bloodstream and reduces the delivery of oxygen to the body's organs and tissues. The health threat from carbon monoxide is most serious for those who suffer from cardiovascular disease, particularly those with angina or peripheral vascular disease. Healthy individuals also are affected but only at higher levels. Exposure to elevated carbon monoxide levels is associated with impairment of visual perception, work capacity, manual dexterity, learning ability and performance of complex tasks.

Source: U.S. Environmental Protection Agency, *National Air Quality and Emissions Trends Report, 1990*, Nov. 1991, pp. 1-8–1-9.

NITROGEN DIOXIDE (NO_2)

AIR CONCENTRATIONS

1981-90: 8 percent decrease (annual mean at 166 sites)

1989-90: 6 percent decrease (annual mean at 211 sites)

EMISSIONS : NO_x

1981-90: 6 percent decrease

1989-90: 1 percent decrease

OVERVIEW

Trends Nitrogen oxide emissions increased 6 percent since 1970 but both emissions (-6 percent) and nitrogen dioxide air quality (-8 percent) showed improvement during the 1980's. The national trend in annual mean NO_2 concentrations was flat for most of the 1980's, however, annual mean NO_2 levels have declined during the past two years. The two primary source categories of nitrogen oxide emissions, and their contribution in 1990, are fuel combustion (57 percent) and transportation (38 percent). The transportation category has decreased 24 percent while fuel combustion emissions are estimated to have increased by 12 percent.

Status In October 1991, EPA designated only one area as nonattainment for NO_2. Los Angeles, CA, which reported an annual mean of 0.056 ppm in 1990, is the only urban area that has recorded violations of the annual NO_2 NAAQS of 0.053 ppm during the past 10 years.

1990 Clean Air Act Although Los Angeles is the only nonattainment area for nitrogen dioxide, the Clean Air Act Amendments of 1990 recognized the need for nitrogen oxide controls due to its contributing role in other problems including ozone (smog) and acid rain. EPA has already issued final tighter tailpipe standards for NO_x as required under the new amendments. Future ozone (smog) control plans will address further NO_x controls and the Acid Rain provisions of the Act calls for a 2 million ton NO_x reduction from affected utilities.

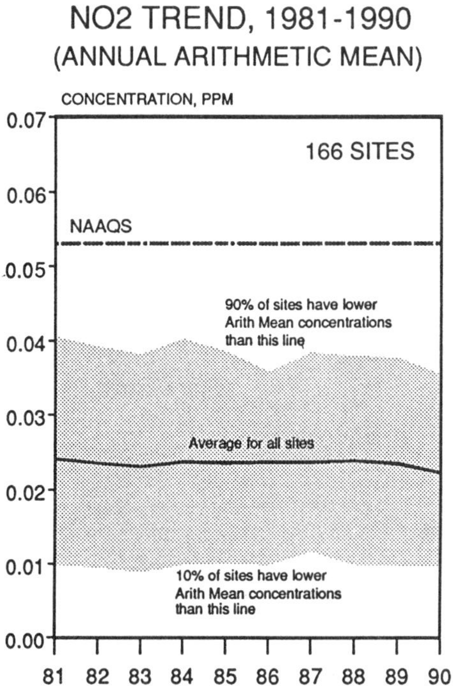

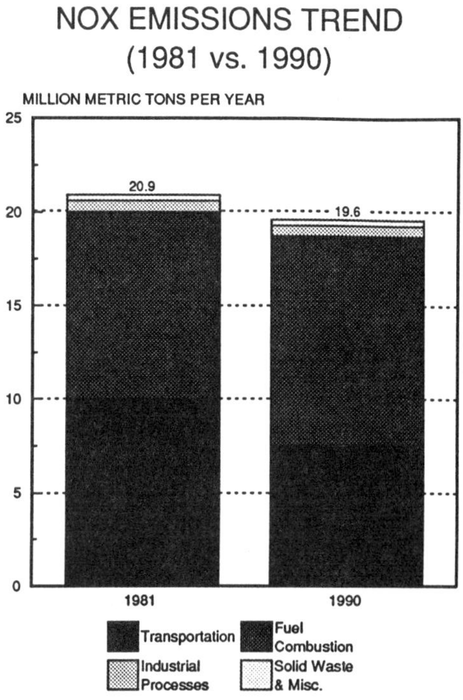

NO_2 EFFECTS

Nitrogen dioxide can irritate the lungs and lower resistance to respiratory infection (such as influenza). The effects of short-term exposure are still unclear but continued or frequent exposure to concentrations higher than those normally found in the ambient air may cause increased incidence of acute respiratory disease in children. Nitrogen oxides are an important precursor both to ozone and to acidic precipitation and may affect both terrestrial and aquatic ecosystems. Atmospheric deposition of NO_x is a potentially significant contributor to ecosystem effects including algal blooms in certain estuaries such as the Chesapeake Bay. In some western areas, NO_x is an important precursor to PM-10 concentrations.

Source: U.S. Environmental Protection Agency, *National Air Quality and Emissions Trends Report, 1990*, Nov. 1991, pp. 1-10–1-11.

OZONE (O_3)

AIR CONCENTRATIONS

1981-90: 10 percent decrease (second highest daily max 1-hour at 471 sites)
51 percent decrease (exceedance days at 471 sites)
1989-90: 1 percent decrease (second highest daily max 1-hour at 590 sites)

EMISSIONS : VOC

1981-90: 12 percent decrease
1989-90: 1 percent increase

OVERVIEW

Trends Ground level ozone, the primary constituent of smog, has been a pervasive pollution problem for the U.S. Ambient trends during the 1980's were influenced by varying meteorological conditions. Relatively high 1983 and 1988 ozone levels are likely attributed in part to hot, dry, stagnant conditions in some areas of the country. Both 1989 and 1990 levels showed improvement but the complexity of the ozone problem warrants caution in interpreting the data. There have been recent control measures, such as lower Reid Vapor Pressure (RVP) for gasoline resulting in lower fuel volatility and lower NO_x and VOC emissions from tailpipes. Emission estimates for volatile organic compounds (VOCs), which contribute to ozone formation, are estimated to have improved by 31 percent since 1970 and 12 percent since 1981. However, these volatile organic compound (VOC) emission estimates represent annual totals. NO_x emissions, the other major precursor factor in ozone formation, decreased 6 percent between 1981 and 1990. While these annual emission totals are the best national numbers now available, ozone is predominantly a warm weather problem and seasonal emission trends would be preferable.

Status In October 1991, EPA designated 98 areas as nonattainment for O_3.

1990 Clean Air Act The Act expanded the framework for designating areas as attainment or nonattainment for ozone by further classifying areas based upon the magnitude of their problem. Ozone nonattainment areas are now classified as marginal, moderate, serious, severe or extreme. This allows more flexibility in the required control program. The Act includes a variety of new requirements for cars and other sources of ozone precursors, including the introduction of cleaner (reformulated) gasoline beginning in 1995 into the nine U.S. cities with the worst ozone problems.

OZONE TREND, 1981-1990
(ANNUAL 2ND DAILY MAX HOUR)

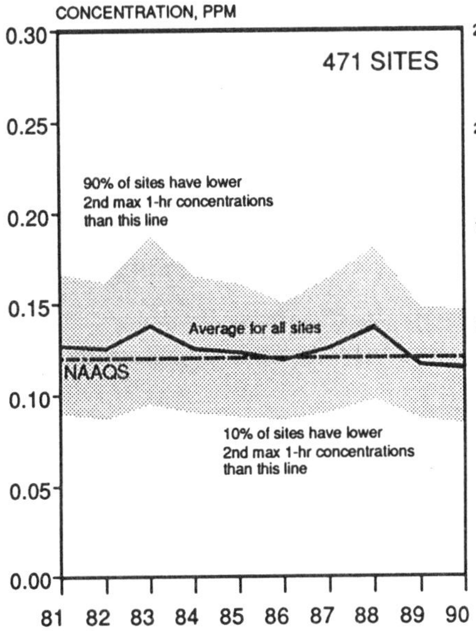

VOC EMISSIONS TREND
(1981 vs. 1990)

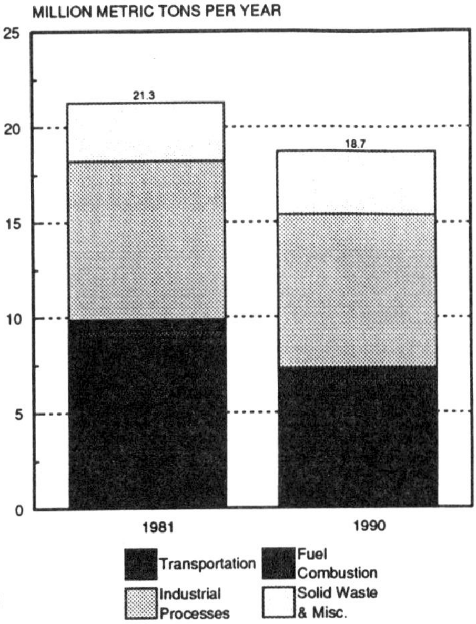

O_3 EFFECTS

The reactivity of ozone causes health problems because it damages biological tissues and cells. Recent scientific evidence indicates that ambient levels of ozone not only affect people with impaired respiratory systems, such as asthmatics, but healthy adults and children, as well. Exposure to ozone for 6 - 7 hours at relatively low concentrations (i.e. 0.08 ppm) has been found to significantly reduce lung function in normal, healthy people during periods of moderate exercise. This decrease in lung function often is accompanied by such symptoms as chest pain, coughing, nausea and pulmonary congestion. Though less well established in humans, animal studies have demonstrated that repeated exposure to ozone for months to years can produce permanent structural damage in the lungs and accelerate the rate of lung function loss and aging of the lungs. Ozone is responsible each year for agricultural crop yield loss in the U.S. of several billion dollars and causes noticeable foliar damage in many crops and species of trees. Forest and ecosystem studies indicate that damage is resulting from current ambient ozone levels.

Source: U.S. Environmental Protection Agency, *National Air Quality and Emissions Trends Report, 1990*, Nov. 1991, pp. 1-12–1-13.

LEAD (Pb)

AIR CONCENTRATIONS

1981-90: 85 percent decrease (maximum quarterly average at 202 sites)

1989-90: 12 percent decrease (maximum quarterly average at 229 sites)

EMISSIONS

1981-90: 87 percent decrease in total lead emissions
(95 percent decrease in lead emissions from transportation sources)

1989-90: 1 percent decrease in total lead emissions
(no change in lead emissions from transportation sources)

OVERVIEW

Trends Total lead emissions have dropped 97 percent since 1970 due principally to reductions in ambient lead levels from automotive sources. Ambient lead (Pb) concentrations in urban areas throughout the country have decreased 85 percent since 1981 while emissions decreased by 87 percent. The drop in Pb consumption and subsequent Pb emissions was brought about by the increased use of unleaded gasoline in catalyst-equipped cars (89 percent of the total gasoline market in 1990) and the reduced Pb content in leaded gasoline.

Status In 1990, the reduction of exposure to lead became a top priority objective for the Agency. Among other things, EPA identified 29 stationary sources with potential problems. An assessment of these sources' compliance status, ambient monitoring availability and State implementation plan (SIP) adequacy was completed.

1990 Clean Air Act The Amendments, for the first time, authorize EPA to designate areas nonattainment, attainment or unclassifiable for the lead NAAQS. As such, EPA has designated as nonattainment 12 areas which have recently recorded violations of the lead NAAQS. EPA has also designated as unclassifiable 9 areas for which existing air quality data are insufficient at this time to designate as either attainment or nonattainment. As States submit designation requests and as ambient monitoring data become available, EPA will proceed to designate additional lead areas as appropriate. Once an area is designated nonattainment for the lead NAAQS, States must submit revised pollution control plans within 18 months of the area's nonattainment designation.

PB TREND, 1981-1990
(ANNUAL MAX QUARTERLY AVG)

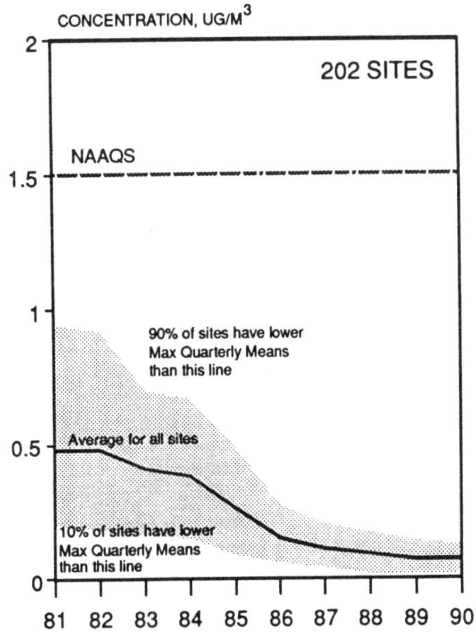

PB EMISSIONS TREND
(1981 vs. 1990)

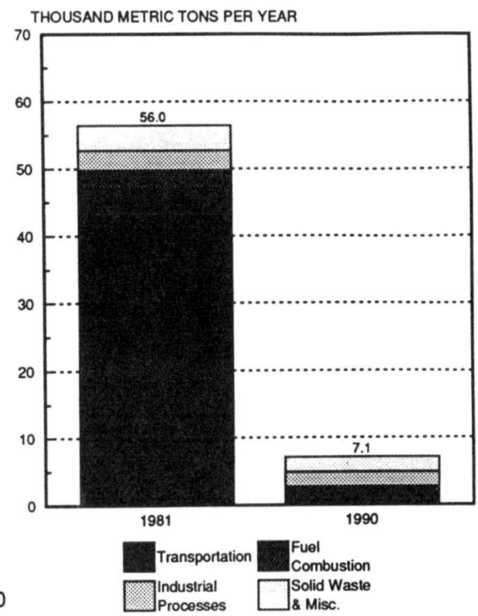

PB EFFECTS

Exposure to lead can occur through multiple pathways, including inhalation of air, diet and ingestion of lead in food, water, soil or dust. Lead accumulates in the body in blood, bone and soft tissue. Because it is not readily excreted, lead also affects the kidneys, liver, nervous system and blood-forming organs. Excessive exposure to lead may cause neurological impairments such as seizures, mental retardation and/or behavioral disorders. Even at low doses, lead exposure is associated with changes in fundamental enzymatic, energy transfer and homeostatic mechanisms in the body. Fetuses, infants and children are especially susceptible to low doses of lead, often suffering central nervous system damage. Recent studies have also shown that lead may be a factor in high blood pressure and subsequent heart disease in middle-aged white males.

Source: U.S. Environmental Protection Agency, *National Air Quality and Emissions Trends Report, 1990*, Nov. 1991, pp. 1-14–1-15.

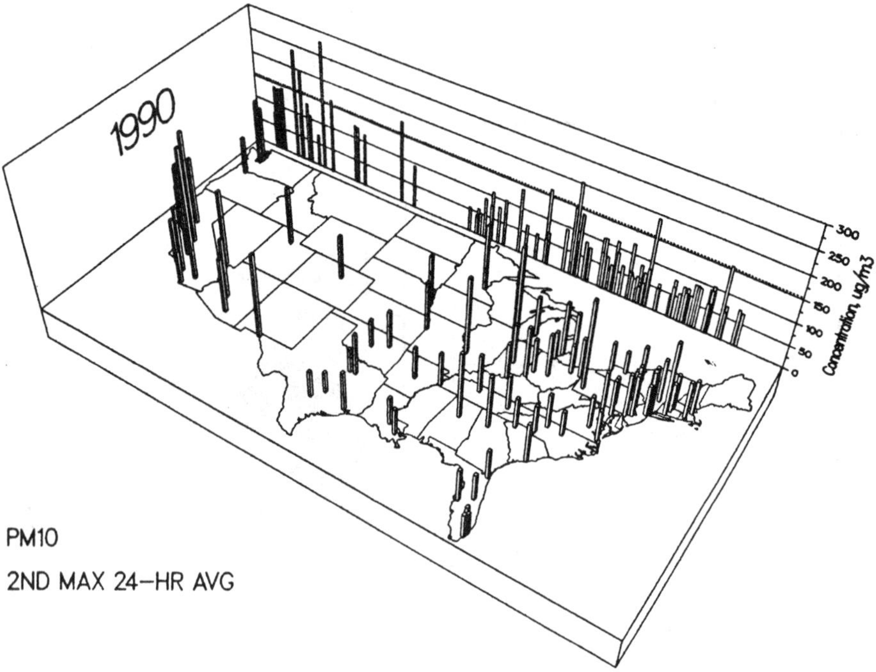

United States map of the highest second maximum 24-hour average PM-10 concentration by MSA, 1990.

The map for PM-10 shows the 1990 highest second maximum 24-hour average PM-10 concentration in metropolitan areas greater than 500,000 population. Concentrations above the level of the 24-hour PM-10 standard of 150 µg/m³ are found in 9 of these metropolitan areas.

Source: U.S. Environmental Protection Agency, *National Air Quality and Emissions Trends Report, 1990*, Nov. 1991, p. 4-12.

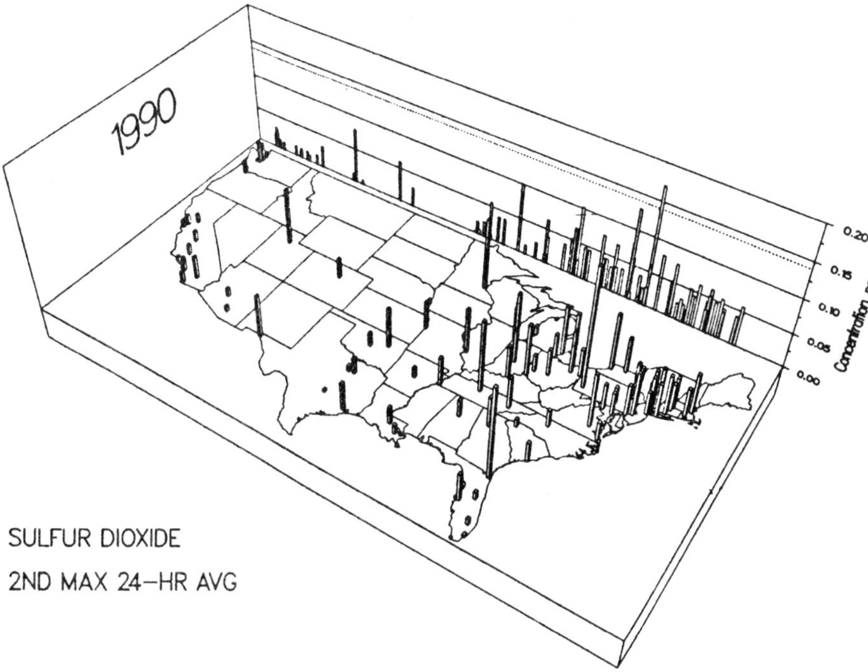

United States map of the highest second maximum 24-hour average sulfur dioxide concentration by MSA, 1990.

The map for sulfur dioxide shows the highest second highest 24-hour average sulfur dioxide concentration by MSA in 1990. Pittsburgh, PA is the only large urban area which had ambient concentrations above the 24-hour NAAQS of 365 µg/m³ (0.14 ppm).

Source: U.S. Environmental Protection Agency, *National Air Quality and Emissions Trends Report, 1990*, Nov. 1991, p. 4-15.

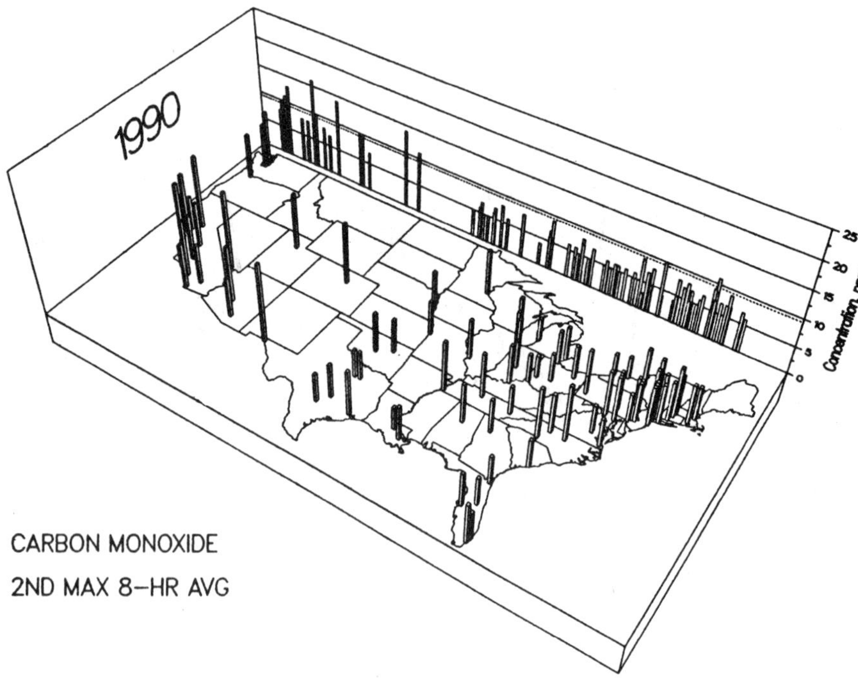

United States map of the highest second maximum nonoverlapping 8-hour average carbon monoxide concentration by MSA, 1990.

The map for carbon monoxide shows the highest second highest 8-hour value recorded in 1990. Twelve of these urban areas have air quality exceeding the 9 ppm level of the standard. The highest concentration recorded in 1990 is found in Los Angeles, CA.

Source: U.S. Environmental Protection Agency, *National Air Quality and Emissions Trends Report, 1990*, Nov. 1991, p. 4-16.

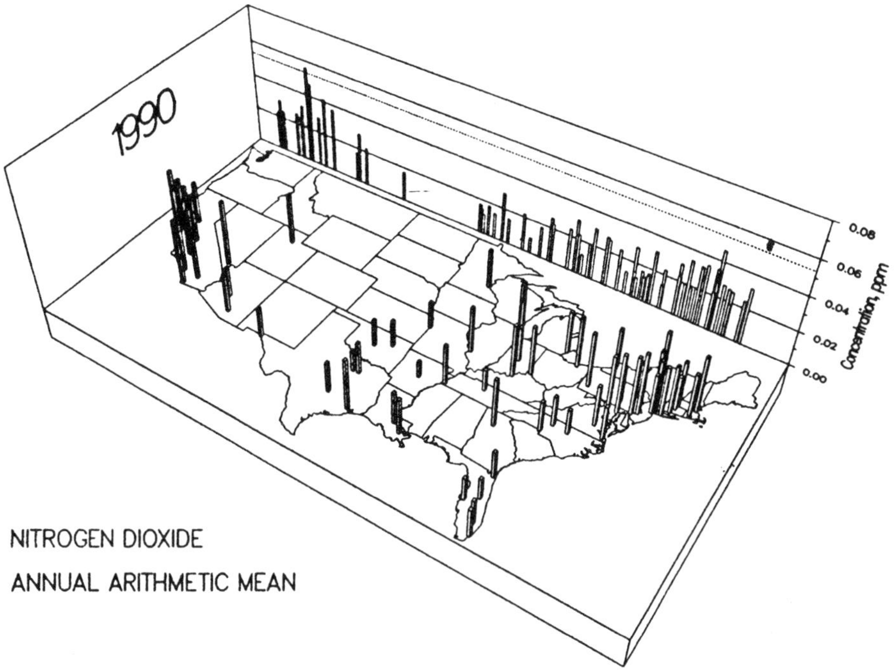

United States map of the highest annual arithmetic mean nitrogen dioxide concentration by MSA, 1990.

The map for nitrogen dioxide displays the maximum annual mean measured in the nation's largest metropolitan areas during 1990. Los Angeles, California, with an annual NO_2 mean of 0.056 ppm is the only area in the country exceeding the NO_2 air quality standard of 0.053 ppm.

Source: U.S. Environmental Protection Agency, *National Air Quality and Emissions Trends Report, 1990*, Nov. 1991, p. 4-17.

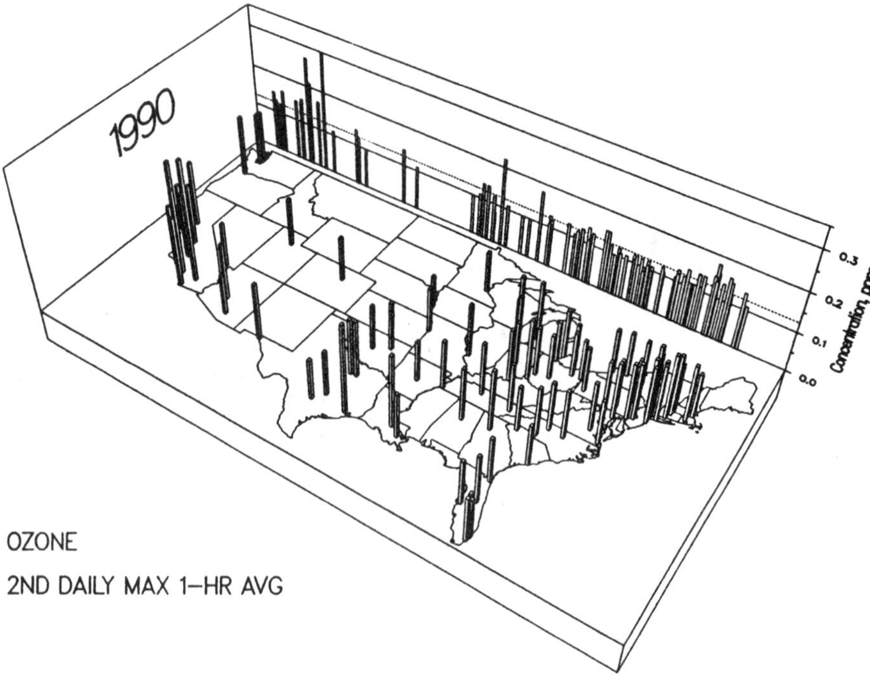

United States map of the highest second daily maximum 1-hour average ozone concentration by MSA, 1990.

The ozone map shows the second highest daily maximum 1-hour concentration in the 90 largest metropolitan areas in the Continental U.S. As shown, 39 of these areas did not meet the 0.12 ppm standard in 1990. The highest concentrations are observed in Southern California, but high levels also persist in the Texas Gulf Coast, Northeast Corridor and other heavily populated regions.

Source: U.S. Environmental Protection Agency, *National Air Quality and Emissions Trends Report, 1990*, Nov. 1991, p. 4-18.

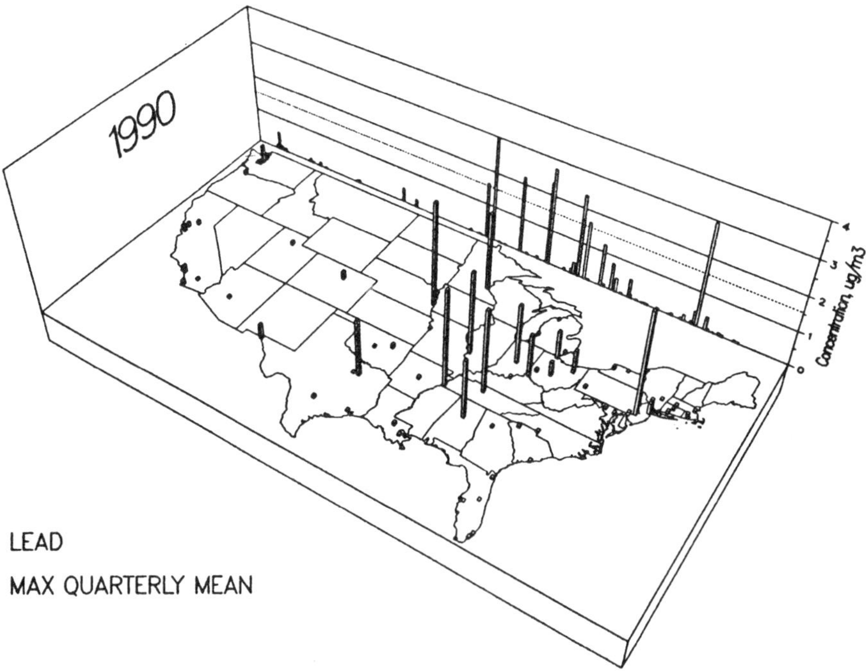

United States map of the highest maximum quarterly average lead concentration by MSA, 1990.

The map for Pb displays maximum quarterly average concentrations in the nation's largest metropolitan areas. Exceedances of the Pb NAAQS are found in nine areas in the vicinity of nonferrous smelters or other point sources of lead. Because of the switch to unleaded gasoline, areas primarily affected by automotive lead emissions show levels below the current standard of 1.5 $\mu g/m^3$.

Source: U.S. Environmental Protection Agency, *National Air Quality and Emissions Trends Report, 1990*, Nov. 1991, p. 4-19.

National Sulfur Oxides Emission Estimates, 1981-1990

	(million metric tons/year)									
SOURCE CATEGORY	1981	1982	1983	1984	1985	1986	1987	1988	1989	1990
Transportation	0.9	0.8	0.8	0.8	0.9	0.9	0.9	0.9	1.0	0.9
Fuel Combustion	17.8	17.3	16.7	17.4	17.0	16.9	16.6	16.6	16.8	17.1
Industrial Processes	3.8	3.1	3.1	3.2	3.2	3.2	3.0	3.1	3.0	3.1
Solid Waste	0.0	0.0	0.0	0.0	0.0	0.0	0.0	0.0	0.0	0.0
Miscellaneous	0.0	0.0	0.0	0.0	0.0	0.0	0.0	0.0	0.0	0.0
TOTAL	22.5	21.2	20.6	21.5	21.1	20.9	20.5	20.6	20.8	21.2

NOTE: The sums of sub-categories may not equal total due to rounding.

National Carbon Monoxide Emission Estimates, 1981-1990

	(million metric tons/year)									
SOURCE CATEGORY	1981	1982	1983	1984	1985	1986	1987	1988	1989	1990
Transportation	55.4	52.9	52.4	50.6	47.9	44.6	43.3	41.2	40.0	37.6
Fuel Combustion	7.7	8.2	8.2	8.3	7.5	7.5	7.6	7.6	7.8	7.5
Industrial Processes	5.9	4.4	4.3	4.7	4.4	4.2	4.3	4.6	4.6	4.7
Solid Waste	2.1	2.0	1.9	1.9	1.9	1.8	1.8	1.7	1.7	1.7
Miscellaneous	6.4	4.9	7.8	6.4	7.1	5.1	6.4	9.5	6.3	8.6
TOTAL	77.5	72.5	74.5	71.9	68.7	63.2	63.4	64.7	60.4	60.1

NOTE: The sums of sub-categories may not equal total due to rounding.

Source: U.S. Environmental Protection Agency, *National Air Quality and Emissions Trends Report, 1990*, Nov. 1991, pp. 3-13–3-18.

National Volatile Organic Compound Emission Estimates, 1981-1990

(million metric tons/year)										
SOURCE CATEGORY	**1981**	**1982**	**1983**	**1984**	**1985**	**1986**	**1987**	**1988**	**1989**	**1990**
Transportation	8.9	8.3	8.2	8.1	7.6	7.2	7.1	6.9	6.4	6.4
Fuel Combustion	1.0	1.0	1.0	1.0	0.9	0.9	0.9	0.9	0.9	0.9
Industrial Processes	8.3	7.5	7.9	8.9	8.5	8.0	8.3	8.1	8.1	8.1
Solid Waste	0.7	0.6	0.6	0.6	0.6	0.6	0.6	0.6	0.6	0.6
Miscellaneous	2.4	2.1	2.7	2.6	2.5	2.2	2.4	2.9	2.5	2.7
TOTAL	21.3	19.6	20.4	21.2	20.1	19.0	19.3	19.4	18.5	18.7

NOTE: The sums of sub-categories may not equal total due to rounding.

National Nitrogen Oxides Emission Estimates, 1981-1990

(million metric tons/year)										
SOURCE CATEGORY	**1981**	**1982**	**1983**	**1984**	**1985**	**1986**	**1987**	**1988**	**1989**	**1990**
Transportation	9.9	9.4	8.9	8.8	8.9	8.3	8.1	8.1	7.9	7.5
Fuel Combustion	10.0	9.8	9.6	10.2	10.2	10.0	10.5	10.9	11.1	11.2
Industrial Processes	0.6	0.5	0.5	0.6	0.6	0.6	0.6	0.6	0.6	0.6
Solid Waste	0.1	0.1	0.1	0.1	0.1	0.1	0.1	0.1	0.1	0.1
Miscellaneous	0.2	0.1	0.2	0.2	0.2	0.2	0.2	0.3	0.2	0.3
TOTAL	20.9	20.0	19.4	19.8	19.9	19.1	19.4	20.0	19.8	19.6

NOTE: The sums of sub-categories may not equal total due to rounding.

Source: U.S. Environmental Protection Agency, *National Air Quality and Emissions Trends Report, 1990*, Nov. 1991, pp. 3-22–3-27.

National Lead Emission Estimates, 1981-1990

(thousand metric tons/year)										
SOURCE CATEGORY	**1981**	**1982**	**1983**	**1984**	**1985**	**1986**	**1987**	**1988**	**1989**	**1990**
Transportation	46.5	47.0	40.8	34.7	14.7	3.5	3.0	2.6	2.2	2.2
Fuel Combustion	2.8	1.7	0.6	0.5	0.5	0.5	0.5	0.5	0.5	0.5
Industrial Processes	3.0	2.7	2.4	2.3	2.3	1.9	1.9	2.0	2.3	2.2
Solid Waste	3.7	3.1	2.7	2.7	2.6	2.6	2.6	2.5	2.3	2.2
Miscellaneous	0.0	0.0	0.0	0.0	0.0	0.0	0.0	0.0	0.0	0.0
TOTAL	56.0	54.5	46.6	40.2	20.1	8.4	8.0	7.6	7.2	7.1

NOTE: The sums of sub-categories may not equal total due to rounding.

National Total Particulate Emission Estimates, 1981-1990

(million metric tons/year)										
SOURCE CATEGORY	**1981**	**1982**	**1983**	**1984**	**1985**	**1986**	**1987**	**1988**	**1989**	**1990**
Transportation	1.3	1.3	1.3	1.3	1.4	1.4	1.4	1.5	1.5	1.5
Fuel Combustion	2.3	2.2	2.0	2.1	1.8	1.8	1.8	1.7	1.8	1.7
Industrial Processes	3.0	2.6	2.4	2.8	2.8	2.6	2.5	2.7	2.7	2.8
Solid Waste	0.4	0.3	0.3	0.3	0.3	0.3	0.3	0.3	0.3	0.3
Miscellaneous	0.9	0.7	1.1	0.9	1.0	0.8	0.9	1.3	0.9	1.2
TOTAL	8.0	7.1	7.1	7.4	7.2	6.7	6.9	7.5	7.2	7.5

NOTE: The sums of sub-categories may not equal total due to rounding.

Source: U.S. Environmental Protection Agency, *National Air Quality and Emissions Trends Report, 1990*, Nov. 1991, pp. 3-32–3-4.

National PM-10 Emission Estimates, 1985-1990

(million metric tons/year)						
SOURCE CATEGORY	**1985**	**1986**	**1987**	**1988**	**1989**	**1990**
Transportation	1.3	1.3	1.3	1.4	1.5	1.5
Fuel Combustion	1.1	1.1	1.1	1.1	1.1	1.1
Industrial Processes	2.7	2.5	2.4	2.6	2.6	2.7
Solid Waste	0.3	0.2	0.2	0.2	0.2	0.2
Miscellaneous	0.7	0.5	0.7	1.0	0.7	0.9
TOTAL	6.0	5.6	5.8	6.3	6.1	6.4

NOTE: The sums of sub-categories may not equal total due to rounding.

Source: U.S. Environmental Protection Agency, *National Air Quality and Emissions Trends Report, 1990*, Nov. 1991, p. 3-8.

APPENDIX B

CLEAN WATER CHARTS

Designated Use Support in Rivers and Streams 312

Percent of Impaired River Miles Affected by Causes of Pollution 313

Percent of Impaired River Miles Affected by Sources of Pollution 313

Impaired River Miles Affected by Sources of Pollution 314

Attainment of Clean Water Goals in Rivers and Streams 316

Designated Use Support in Lakes and Reservoirs 317

Impaired Lake Acres Affected by Sources of Pollution 318

Attainment of Clean Water Goals in Lakes and Reservoirs 320

Designated Use Support in Estuaries 321

Impaired Estuary Square Miles Affected by Causes of Pollution 321

Percent of Impaired Estuary Square Miles Affected by Causes of Pollution 322

Percent of Impaired Estuary Square Miles Affected by Sources of Pollution 322

Attainment of Clean Water Act Goals in Assessed Estuaries 323

State Attainment of Clean Water Act Goals in Estuaries 323

Designated Use Support in Oceans 324

State Attainment of Clean Water Act Goals in Oceans 324

Wetlands Losses in the United States, 1780s to 1980s 325

Most Frequently Observed Ground-Water Contaminants by Number of States and Territories Reporting 326

Designated Use Support in Rivers and Streams

State	Total River Miles	Miles Assessed: Total	Miles Assessed: Percent Evaluated	Miles Assessed: Percent Monitored	Miles Fully Supporting	Miles Threatened	Miles Partially Supporting	Miles Not Supporting
Alabama	40,600	12,016	69	31	8,783	159	2,003	1,071
Arizona	17,537	5,185	72	28	206	1,120	2,142	1,717
Arkansas	11,310	4,713	54	45	1,636	—	539	2,538
California	26,970	12,122	30	70	4,612	—	5,920	1,590
Colorado	31,470	30,582	72	28	27,662	974	1,221	725
Connecticut	8,400	893	40	60	420	190	229	54
Delaware	643	643	11	89	132	15	94	402
Delaware River Basin	206	206	—	—	0	199	0	7
District of Columbia	40	38	0	100	0	0	3	35
Florida	12,659	7,908	37	63	4,890	258	1,863	897
Georgia	20,000	20,000	66	34	19,395	—	556	49
Hawaii	349	349	28	72	265	0	80	4
Illinois	14,080	13,122	20	80	5,522	179	7,263	158
Indiana	20,000	4,917	46	54	2,570	416	1,143	788
Iowa	18,300	7,156	94	6	0	90	6,552	514
Kansas	13,295	12,144	1	99	345	0	1,078	10,721
Kentucky	18,490	9,556	55	45	6,429	130	1,268	1,729
Louisiana	14,180*	13,101	65	35	3,290	5,420	3,053	1,338
Maine	31,672	31,672	75	25	31,107	0	0	565
Maryland	17,000	17,000	85	15	15,657	0	1,218	125
Massachusetts	8,053	1,624	15	85	475	94	467	588
Michigan	36,350	36,350	37	63	34,989	—	—	1,361
Minnesota	91,944	5,316	12	88	1,425	—	517	3,374
Mississippi	15,839	15,839	87	13	7,663	480	7366	330
Missouri	21,064	21,064	65	35	11,220	—	9827	17
Montana	51,212	51,212	95	5	34,808	3,194	11,944	1,266
Nebraska	24,000	7,331	8	92	4,025	—	1,914	1,392
Nevada	7,500	1,447	4	96	405	45	362	635
New Hampshire	14,544	1,348	36	64	20	839	339	150
New Mexico	3,500	3,125	20	80	0	8	2,843	274
New York	70,000	70,000	95	5	63,771	1,439	3,582	1,208
North Carolina	37,222	35,234	67	33	12,766	10,928	9,440	2,100
North Dakota	11,868	9,173	54	46	180	6,709	2,284	0
Ohio	43,917	7,444	—	—	2,231	437	1,540	3,236
Ohio River Valley	981	981	21	79	0	0	981	0
Oklahoma	19,791	4,400	55	45	269	2,305	1,826	0
Oregon	90,000	27,738	87	13	12,546	—	8,497	6,695
Pennsylvania	50,000	23,832	53	47	19,137	—	1,982	2,713
Puerto Rico	5,373	5,373	80	20	2,078	567	1,207	1,521
Rhode Island	724	626	10	90	201	287	29	109
South Carolina	9,900	3,493	0	100	2,758	0	275	460
South Dakota	9,937	4,028	17	83	673	551	1,764	1,040
Tennessee	19,124	10,247	62	38	4,652	1,872	2,620	1,103
Texas	80,000	16,184	3	97	13,548	0	0	2,636
Utah	11,779	8,874	0	100	1,544	238	5,658	1,434
Vermont	5,266	5,266	91	9	3,140	1,194	720	212
Virginia	27,240	10,809	46	54	8,862	—	1,468	479
Washington	40,492	4,897	13	87	1,548	562	1,319	1,468
West Virginia	28,361	17,456	62	38	7,160	0	9,180	1,116
Wisconsin	43,600	13,595	80	20	6,878	1,504	4,946	267
Wyoming	19,437	19,437	67	33	15,269	811	3,350	7
Totals	1,216,219	647,066	63	37	407,162	43,214	134,472	62,218
Percent of Assessed Waters					62.9%	6.7%	20.8%	9.6%

— Not reported.

*U.S. EPA Reach File Version 1 estimate (1982).

Source: U.S. Environmental Protection Agency, *National Water Quality Inventory*, 1990 Report to Congress, Apr. 1992, p. 4.

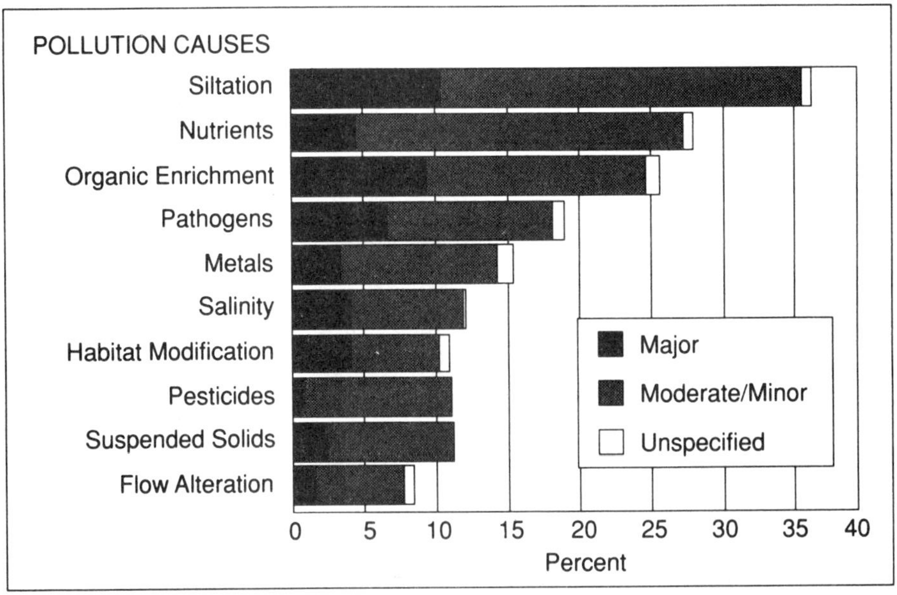

Percent of Impaired River Miles Affected by Causes of Pollution

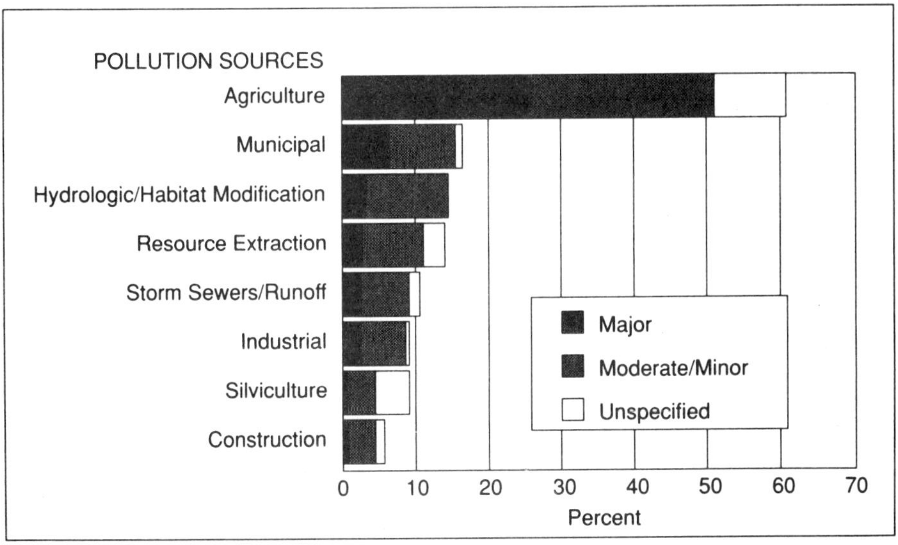

Percent of Impaired River Miles Affected by Sources of Pollution

Source: U.S. Environmental Protection Agency, *National Water Quality Inventory*, 1990 Report to Congress, Apr. 1992, pp. 9–10.

Impaired River Miles Affected by Sources of Pollution

State	Total Impaired Waters*	Agriculture		Municipal		Hydrologic/ Habitat Mod		Resource Extraction		Storm Sewers/ Runoff	
		Major	Mod/Min	Major	Mod/Min	Major	Mod/Min	Major	Mod/Min	Major	Mod/Min
Alabama	3,074	554	502	388	192	80	125	513	155	12	261
Arizona	3,859	—	3,034	—	670	—	1,264	—	1,847	—	396
Arkansas**	3,077	2,597	—	182	—	—	—	200	—	15	—
Connecticut	283	—	47	133	113	6	50	—	42	1	136
Delaware	496	78	352	13	61	—	—	—	—	9	163
Delaware River Basin	7	—	—	7	—	—	—	—	—	—	—
District of Columbia	38	—	—	1	—	—	3	—	—	16	22
Florida	2,760	188	1,648	35	642	20	887	—	438	84	1,779
Georgia	605	—	—	41	54	—	—	—	—	3	385
Illinois	7,421	2,065	—	405	—	202	—	20	—	9	—
Iowa	7,066	6,825	213	170	406	4	29	2	85	172	1,082
Indiana	1,931	1,338	208	646	1,030	—	—	29	—	7	—
Kansas	11,799	7,725	3,261	949	298	—	—	118	64	664	24
Kentucky	2,997	1,046	185	1,151	25	153	—	833	34	219	42
Louisiana	4,391	1,182	1,355	212	2,624	123	2,823	111	3,400	196	1,472
Maine	565	72	47	8	63	—	4	—	1	3	12
Maryland	1,343	22	912	—	—	2	1	87	45	—	—
Massachusetts	1,055	11	60	138	229	—	—	—	—	288	538
Michigan	1,361	203	118	459	134	189	31	6	—	182	262
Mississippi	7,696	723	6,436	128	916	—	19	—	230	57	208
Missouri	9,844	7,137	—	10	36	1	3,669	15	40	188	—
Montana	13,210	502	10,671	46	1,077	300	5,198	239	2,158	—	106
Nebraska	3,306	2,423	526	605	831	333	53	—	—	189	528
Nevada	997	441	218	—	—	175	247	101	—	—	139
New Hampshire	489	—	—	140	131	—	—	—	—	—	—
New Mexico	3,117	2,403	345	55	288	93	620	145	179	33	4
New York	4,790	33	28	160	156	97	41	1	24	10	113
North Carolina**	11,540	5,502	—	172	80	—	—	—	—	922	—
Ohio	4,776	635	1,415	1,945	732	645	1,037	594	265	275	885
Oregon**	15,192	7,605	—	1,062	—	—	—	2,280	—	1,675	—
Pennsylvania**	4,695	624	—	432	—	30	—	2,548	—	57	—
Puerto Rico	2,728	172	608	60	60	46	—	—	14	312	119
Rhode Island	138	36	5	21	31	—	13	—	—	36	83
South Carolina	735	282	161	194	46	—	6	—	—	93	110
South Dakota	2,804	1,275	1,097	11	43	—	—	—	62	47	215
Tennessee	3,723	1,590	1,515	511	1,028	1,312	1,328	350	763	173	822
Texas	2,636	5	44	1,360	110	18	—	—	—	94	—
Utah	7,092	1,306	4,562	65	509	216	—	293	—	—	—
Vermont	932	227	239	39	135	195	210	51	35	71	95
Washington	2,787	576	549	152	1,250	715	595	83	77	650	216
West Virginia	10,296	390	2,359	323	1,446	144	585	1,019	3,595	300	558
Wyoming	3,357	2,192	734	6	110	569	378	54	770	10	282
Totals	171,008	59,985	43,454	12,435	15,559	5,668	19,216	9,692	14,323	7,072	11,057
Combined Totals			103,439		27,994		24,884		24,015		18,129
Percent of Impaired Waters			60.5%		16.4%		14.6%		14.0%		10.6%

* The sum of partially and nonsupporting river miles (Table 1-1).

** These States did not specify the degree of impact (i.e., Major or Moderate/Minor); river miles were placed in the "Major" column for national reporting purposes.

— Zero or not reported.

Industrial		Silviculture		Construction		Unknown		Land Disposal		Combined Sewers	
Major	Mod/Min	Major	Mod/Min	Major	Mod/Min	Major	Mod/Min	Major	Mod/Min	Major	Mod/Min
70	130	64	132	15	102	22	—	—	35	—	23
—	242	—	166	—	527	—	1,303	—	617	—	—
124	—	261	—	15	—	—	—	81	—	—	—
70	22	—	—	—	42	—	6	9	91	107	11
22	55	—	—	—	—	—	—	—	—	—	16
—	—	—	—	—	—	—	—	—	—	—	—
1	—	—	—	—	1	—	—	—	6	11	1
104	430	—	142	21	823	119	—	15	1,027	—	—
5	117	—	—	—	—	—	—	—	—	—	—
24	—	—	—	—	—	—	—	14	—	103	—
78	31	—	—	—	37	—	—	2	117	—	—
208	73	—	—	—	—	—	—	42	—	625	83
—	17	—	—	—	—	6	—	—	—	—	—
182	30	—	—	2	—	204	—	75	50	—	—
368	2,325	17	1,322	7	528	—	1,627	22	855	20	169
103	197	—	—	—	7	—	1	43	13	1	21
—	—	—	—	21	2	—	—	245	74	—	—
87	50	—	—	—	29	—	—	71	90	187	51
80	157	5	8	78	16	197	127	18	3	317	62
136	834	—	405	—	—	—	286	5	58	—	—
4	—	—	—	—	—	—	—	1	—	—	—
—	339	—	1,690	—	1,267	—	—	—	340	—	—
186	208	—	—	—	—	334	252	—	—	—	—
—	—	—	40	—	—	59	—	22	—	—	—
29	—	—	—	—	—	—	11	2	6	40	17
—	—	34	130	25	205	66	314	10	99	—	—
28	90	—	38	—	41	94	16	62	141	36	78
102	1,226	—	—	564	—	—	—	—	—	—	—
745	515	6	12	7	62	232	79	144	526	102	60
368	—	7,580	—	1,420	—	—	—	—	—	—	—
250	—	—	—	—	—	110	—	103	—	32	—
37	124	—	—	17	23	—	16	203	527	2	—
—	74	—	—	15	18	19	—	33	39	6	6
72	26	—	—	1	—	216	26	—	—	—	—
11	—	—	—	—	26	—	—	—	26	—	—
218	353	76	66	153	928	18	62	14	187	62	70
28	—	—	—	—	—	26	681	44	—	—	—
65	509	161	—	—	—	—	—	—	—	—	—
10	56	22	66	137	73	—	2	6	132	—	4
165	476	171	136	414	105	135	846	262	357	1	39
966	1,413	198	2,388	209	607	650	1,104	56	168	190	283
334	169	58	65	362	858	—	—	—	—	—	—
5,280	10,288	8,653	6,806	3,483	6,327	2,507	6,759	1,604	5,584	1,842	994
	15,568		15,459		9,810		9,266		7,188		2,836
	9.1%		9.0%		5.7%		5.4%		4.2%		1.7%

Source: U.S. Environmental Protection Agency, *National Water Quality Inventory*, 1990 Report to Congress, Apr. 1992, pp. 12–13.

Attainment of Clean Water Act Goals in Rivers and Streams

State	Fishable Goal (miles)					Swimmable Goal (miles)				
	Assessed	Meeting	Partially Meeting	Not Meeting	Not Attainable	Assessed	Meeting	Partially Meeting	Not Meeting	Not Attainable
Alabama	11,857	8,703	1,669	1,049	436	11,857	8,703	1,669	1,049	436
Arizona	5,296	662	2,312	2,061	260	5,296	1,810	566	645	2,275
Arkansas	11,310	9,426	1,535	349	—	11,310	9,236	0	2,074	—
California	11,448	9,069	1,492	887	—	10,463	7,947	1,620	896	—
Colorado	28,770	28,105	—	572	93	31,377	9,062	—	254	22,061
Connecticut	893	664	170	57	2	893	679	94	118	2
Delaware	643	498	71	74	0	643	98	56	365	124
Delaware River Basin	206	206	0	0	0	206	199	0	7	0
District of Columbia	39	10	3	26	0	38	0	3	35	0
Florida	7,950	6,750	268	897	35	7,950	6,750	268	897	35
Georgia	20,000	19,395	556	49	0	0	—	—	—	—
Hawaii	349	349	—	0	0	349	349	—	0	0
Illinois	13,123	11,476	1,101	546	0	4,525	1,144	2,003	424	954
Indiana	4,994	2,986	1,143	788	77	2,304	138	153	2,013	0
Iowa	7,155	90	6,620	428	17	7,155	158	1,456	223	5,318
Kansas	12,079	8,838	2,265	976	0	11,942	627	1,546	9,769	0
Kentucky	9,339	6,914	1,702	723	—	3,595	1,481	576	1,538	—
Louisiana	8,535	6,510	1,553	472	—	8,535	5,670	2,055	810	—
Maine	31,672	31,282	0	390	0	31,672	31,506	0	166	0
Maryland	17,000	15,618	1,281	101	0	17,000	16,998	0	2	0
Massachusetts	1,623	1,061	208	354	0	1,624	682	620	319	3
Michigan	36,350	35,632	—	718	0	36,350	36,086	—	264	0
Minnesota	6,079	2,292	681	2,306	800	5021	2,045	548	1,628	800
Mississippi	15,839	8,023	7,350	350	116	15,839	13,086	2,126	511	116
Missouri	21,063	11,216	9,830	17	0	21,069	5,370	—	0	15,699
Montana	51,212	38,474	11,565	1,173	0	51,211	46,096	4,536	579	0
Nebraska	7,330	5,131	1,773	426	0	12,011	1,035	606	1,034	9,336
Nevada	1,559	793	583	83	100	1,574	735	560	57	222
New Hampshire	1,347	1,294	3	50	0	1,390	859	339	150	42
New Jersey	1,719	1,315	307	97	—	592	91	48	453	—
New Mexico	3,117	2,851	—	266	0	3,117	3,117	—	0	0
New York	70,000	69,300	—	0	700	70,000	69,200	—	800	0
North Carolina	37,293	23,820	12,551	922	—	37,293	23,820	12,551	922	—
North Dakota	9,204	8,548	656	0	—	9,204	8,489	715	0	—
Ohio	7,688	2,524	1,482	2,868	814	0	—	—	—	—
Ohio River Valley	981	434	547	0	0	981	579	64	338	0
Oklahoma	3,778	2,161	1,258	359	0	3,633	2,148	412	515	558
Oregon	27,739	26,197	—	1,542	—	27,739	26,773	—	966	—
Pennsylvania	23,832	19,137	1,982	2,713	—	23,832	19,137	1,982	2,713	—
Puerto Rico	5,373	2,757	2,048	566	2	5,373	2,738	1,737	483	415
Rhode Island	626	512	0	24	90	626	512	0	24	90
South Carolina	3,594	3,230	54	310	0	3,438	2,010	388	1,040	0
South Dakota	3,964	3,085	—	879	0	802	396	—	406	0
Tennessee	11,081	10,857	—	224	—	11,081	10,420	—	661	—
Texas	16,203	16,044	—	159	—	16,203	14,435	—	1,768	—
Utah	11,779	1,303	10,476	0	0	4,320	120	4,164	36	0
Vermont	5,266	4,468	608	190	0	5,265	4,854	13	140	258
Virginia	10,809	8,862	1,468	479	—	10,809	8,862	1,468	479	—
Washington	5,141	2,873	1,084	1,184	—	4,928	3,441	311	1,176	—
West Virginia	19,818	17,969	506	1,343	0	19,812	18,415	0	1,397	0
Wisconsin	13,302	8,272	4,334	220	476	13,192	8,235	4,269	212	476
Wyoming	19,437	19,430	—	7	0	947	947	0	0	0
Totals	656,804	527,416	95,095	30,274	4,018	586,386	437,288	49,522	40,356	59,220
Percent of Assessed Waters		80.3%	14.5%	4.6%	0.6%		74.6%	8.4%	6.9%	10.1%

— Not reported.

Source: U.S. Environmental Protection Agency, *National Water Quality Inventory*, 1990 Report to Congress, Apr. 1992, p. 14.

Designated Use Support in Lakes and Reservoirs

State	Number of Lakes	Acres of Lakes	Acres Assessed: Total	Acres Assessed: Percent Evaluated	Acres Assessed: Percent Monitored	Acres Fully Supporting	Acres Threatened	Acres Partially Supporting	Acres Not Supporting
Alabama	43	504,721	504,721	0	100	296,284	147,035	300	61,102
Arizona	182	127,372	121,388	64	36	1,195	92,052	22,343	5,798
Arkansas	77	355,063	355,063	0	100	355,063	—	0	0
California	4,955	1,417,540	1,159,160	19	81	170,652	—	357,130	631,378
Colorado	4,069	265,982	145,305	86	14	132,057	12,276	972	0
Connecticut	6,000	82,900	23,179	33	67	1,858	15,369	5,952	0
Delaware	63	4,499	2,541	31	69	1,263	30	811	437
District of Columbia	8	376	238	0	100	0	0	108	130
Florida	7,712	2,085,120	957,120	17	83	302,336	9,856	533,888	111,040
Georgia	175	417,730	417,730	0	100	404,704	—	13,000	26
Illinois	85,305	305,847	209,036	60	40	595	22,997	101,457	83,987
Indiana	575	106,203	97,536	65	35	97,372	—	63	101
Iowa	264	80,670	80,306	92	8	7,075	35,446	36,820	965
Kansas	235	173,602	170,602	9	91	0	37,598	114,011	18,993
Kentucky	99	228,385	214,861	0	100	100,910	94,839	15,362	3,750
Louisiana	101	866,646	854,331	69	31	548,841	146,858	158,632	0
Maine	5,785	994,560	960,418	22	78	714,028	57,824	188,566	0
Maryland	190	32,583	21,002	100	0	11,463	5,791	3,640	108
Massachusetts	2,871	151,173	64,019	27	73	4,281	15,278	17,803	26,657
Michigan	35,000	840,960	489,433	5	95	50,168	391,904	—	47,361
Minnesota	12,034	3,411,200	2,849,689	22	78	687,064	—	49,817	2,112,808
Mississippi	—	500,000	500,000	51	49	280,389	58,969	160,596	46
Missouri	362	287,543	287,543	15	85	282,905	—	4,564	74
Montana	3,956	740,086	632,705	21	79	31,313	137,627	449,114	14,651
Nebraska	415	145,800	87,267	7	93	60,764	12,245	14,258	0
Nevada	384	425,400	198,725	72	28	129,642	0	54,283	14,800
New Hampshire	1,300	152,788	152,788	71	29	133,943	6,826	7,992	4,027
New Mexico	175	132,690	115,959	15	85	0	—	115,959	0
New York	7,500	750,000	668,000	0	100	377,892	26,942	238,343	24,823
North Carolina	144	305,367	305,355	2	98	195,562	97,442	1,940	10,411
North Dakota	216	619,088	611,075	20	80	13,127	343,132	33,082	221,734
Oklahoma	224,948	1,041,884	441,092	49	51	33,653	193,308	214,131	0
Oregon	6,095	610,808	504,928	85	15	374,303	—	58,918	71,707
Puerto Rico	38	11,146	11,146	20	80	3,433	112	3,598	4,003
Rhode Island	117	16,683	16,683	17	83	1,860	13,592	301	930
South Carolina	1,418	525,000	217,368	0	100	133,180	0	71,068	13,120
South Dakota	799	1,598,285	682,033	2	98	0	548,000	6,885	127,148
Tennessee	117	538,504	538,322	61	39	375,808	53,707	59,456	49,351
Texas	5,700	3,065,600	1,543,898	3	97	1,351,278	7,263	0	185,357
Utah	3,008	481,638	134,201	0	100	0	0	134,201	0
Vermont	604	229,146	227,755	3	97	19,403	158,772	38,611	10,969
Virginia	247	161,290	160,640	70	30	143,856	—	16,784	0
Washington	808	613,582	175,184	1	99	20,112	34,959	39,358	80,755
West Virginia	17,800	23,460	18,870	0	100	8,171	1,775	7,194	1,730
Wisconsin	15,000	957,288	132,202	56	44	9,182	33,102	89,918	0
Wyoming	2,629	427,219	427,219	85	15	306,932	89,883	30,404	0
Totals	459,523	26,813,427	18,488,636	74	26	8,173,917	2,902,809	3,471,633	3,940,277
Percent of Assessed Waters						44.2%	15.7%	18.8%	21.3%

— Not reported.

Source: U.S. Environmental Protection Agency, *National Water Quality Inventory*, 1990 Report to Congress, Apr. 1992, p. 18.

Impaired Lake Acres Affected by Sources of Pollution

State	Total Impaired Waters*	Agriculture		Hydrologic/ Habitat Mod		Storm Sewers/ Runoff		Land Disposal		Municipal	
		Major	Mod/Min	Major	Mod/Min	Major	Mod/Min	Major	Mod/Min	Major	Mod/Min
Arizona	28,141	—	12,474	—	—	—	805	—	390	—	622
Connecticut	5,952	109	3,203	88	916	—	—	23	638	3,203	80
Delaware	1,248	162	812	—	—	186	321	—	—	—	—
District of Columbia	238	—	—	—	103	103	27	—	—	—	—
Florida	644,928	2,304	622,912	—	446,400	576	638,841	—	591,488	1,216	252,480
Georgia	13,026	—	—	—	—	6	1,690	—	—	20	11,310
Illinois	185,444	132,647	52,734	32,018	141,445	3,621	49,984	1,871	33,006	4,925	87,023
Indiana	164	12	—	—	40	—	—	—	—	59	63
Iowa	37,785	20,881	15,421	220	—	16,535	6,890	7	—	12	10,400
Kansas	133,004	607	46,774	—	5,824	8,042	2,334	—	—	—	—
Kentucky	19,112	8,182	—	—	—	—	—	169	—	6,041	455
Maine	188,566	1,794	58,722	—	7,865	31,613	67,476	—	1,999	—	4,845
Maryland	3,748	—	—	—	—	—	104	—	—	—	—
Massachusetts	44,460	272	354	—	—	1,332	1,022	1,337	3,008	1,031	—
Michigan	47,361	28	—	—	—	—	906	—	—	5,279	—
Missouri	4,638	3,778	—	1,730	—	1,270	—	—	—	65	—
Montana	463,765	14,299	364,149	—	299,629	—	546	—	9,708	—	5,100
Nevada	69,083	14,800	6,100	40,005	950	—	—	—	—	6,400	—
New Hampshire	12,019	20	—	25	—	74	34	—	—	3,490	142
New Mexico	115,959	99,037	1,535	—	196	5	—	—	98	—	—
New York	263,166	33,462	93,991	33,877	—	1,347	3,720	2,844	50,086	15,302	8,816
North Carolina**	12,351	445	—	—	—	—	—	—	—	815	—
North Dakota	254,816	4,354	12,830	221,568	579	—	22	—	—	—	2,701
Oregon**	130,625	98,145	—	—	—	10,866	—	13,129	—	485	—
Puerto Rico	7,601	826	1,825	—	—	621	—	—	828	—	—
Rhode Island	1,231	47	—	—	25	844	167	323	109	—	—
South Carolina	84,188	4,570	—	—	—	25	—	—	—	20,450	—
South Dakota	134,033	118,455	7,016	1,339	7,868	90	—	13,560	9,272	—	—
Tennessee	108,807	15,500	50,213	4,789	31,559	20	19,081	—	7,800	14,600	9,556
Texas	185,357	3,624	—	—	—	—	—	—	35	6,640	5
Utah	134,201	15,128	6,518	98,234	3,530	96,932	—	99,715	5,428	97,584	21
Virginia	16,784	—	7,291	—	—	—	4,935	—	21	—	548
Washington	120,113	—	—	—	—	—	—	—	—	—	—
West Virginia	8,924	48	2,184	—	—	39	—	—	—	—	3,434
Wyoming	30,404	28,513	7,665	—	27,005	—	25	—	—	—	8,300
Totals	3,511,242	622,049	1,374,723	433,893	973,934	174,147	798,930	132,978	713,914	187,617	405,901
Combined Totals			1,996,772		1,407,827		973,077		846,892		593,518
Percent of Impaired Waters			56.9%		40.1%		27.7%		24.1%		16.9%

* The sum of partially and nonsupporting lake acres (Table 2-1).

** These States did not specify the degree of impact (i.e., Major or Moderate/Minor); lake acres were placed in the "Major" column for national reporting purposes.

— Zero or not reported.

Unknown		Industrial		Resource Extraction		Construction		Silviculture		Combined Sewers	
Major	Mod/Min	Major	Mod/Min	Major	Mod/Min	Major	Mod/Min	Major	Mod/Min	Major	Mod/Min
—	26,543	—	350	—	6,832	—	152	—	1,084	—	—
—	—	—	1,900	23	—	—	975	—	—	—	—
158	—	—	—	—	—	—	—	—	—	—	—
—	—	—	—	—	—	—	—	—	—	—	—
—	—	—	64,320	—	31,872	—	42,176	—	—	—	—
—	—	—	—	—	—	—	—	—	—	—	—
—	6,602	2,636	8,891	1,088	127,811	26	5,515	75	229	—	25
—	—	—	—	30	—	—	—	—	—	22	23
—	—	—	—	—	—	71	—	1	—	—	—
—	—	—	—	190	50	—	—	—	—	—	—
—	—	—	—	4,862	—	—	—	—	—	—	—
73,007	18,432	—	4,288	—	—	32	1,344	3,442	45,831	—	—
—	—	—	—	—	952	—	—	—	—	—	—
11,460	28,205	45	—	—	—	—	—	—	—	—	—
34,236	2,588	1,780	—	—	2,659	—	—	—	—	1	—
—	—	—	—	—	—	—	—	—	—	—	—
—	—	—	—	—	1,600	—	15,122	—	36,292	—	—
—	—	—	—	20,900	—	—	—	—	—	—	—
83	4,562	3,490	—	—	—	—	—	—	—	—	—
—	19,250	—	—	—	—	—	6,136	244	2,475	—	—
101,094	10,730	—	7,800	—	—	—	35	—	3,180	2,944	—
5,138	—	2,618	—	—	—	—	—	—	—	—	—
—	—	—	—	—	—	—	—	—	—	—	—
—	—	895	—	—	—	—	—	2,755	—	—	—
—	672	—	291	—	—	—	—	—	—	—	—
—	—	—	—	—	—	—	—	—	—	—	—
59,143	—	—	—	—	—	—	—	—	—	—	—
—	—	—	—	—	—	—	—	162	1,698	—	—
—	—	—	40,391	1,924	—	—	10,950	—	4,550	—	—
500	543	—	—	—	—	—	—	—	—	—	—
—	--	96,900	21	2,815	684	2,826	8,532	671	—	—	—
—	—	—	630	—	—	—	875	—	2,028	—	—
41	—	80,000	—	—	80,000	—	—	—	—	—	—
86	7	1,200	—	2,711	4,875	111	14	149	1,636	—	—
—	—	—	—	—	9,520	—	11,506	—	—	—	—
284,946	118,134	189,564	128,882	34,543	266,855	3,066	103,332	7,499	99,003	2,967	48
	403,080		318,446		301,398		106,398		106,502		3,015
	11.5%		9.1%		8.6%		3.0%		3.0%		0.1%

Source: U.S. Environmental Protection Agency, *National Water Quality Inventory*, 1990 Report to Congress, Apr. 1992, pp. 24–25.

Attainment of Clean Water Act Goals in Lakes and Reservoirs

State	Fishable Goal (acres)					Swimmable Goal (acres)				
	Assessed	Meeting	Partially Meeting	Not Meeting	Not Attainable	Assessed	Meeting	Partially Meeting	Not Meeting	Not Attainable
Alabama	488,426	286,039	133,635	63,548	5,204	488,426	286,039	133,635	63,548	5,204
Arizona	121,394	17,747	96,224	7,398	25	121,376	113,179	6,742	120	1,335
Arkansas	355,063	355,063	0	0	—	355,063	355,063	0	0	—
California	1,038,845	538,832	54,931	445,082	—	1,037,225	619,089	70,519	347,617	—
Colorado	148,396	146,494	—	1,862	40	145,055	112,539	—	0	32,516
Connecticut	23,179	20,304	2,875	0	0	23,179	23,179	0	0	0
Delaware	2,542	2,384	0	158	0	2,543	1,294	969	280	0
District of Columbia	238	108	27	103	0	238	0	0	238	0
Florida	957,120	845,696	384	111,040	0	957,120	845,696	384	111,040	0
Georgia	417,730	404,704	13,000	26	0	0	—	—	—	—
Illinois	209,036	160,402	20,920	27,714	—	209,035	136,436	59,765	12,834	—
Indiana	97,536	97,372	63	101	0	97,524	97,372	63	89	0
Iowa	80,306	42,856	26,332	11,115	3	80,306	42,787	29,706	5,586	2,227
Kansas	173,602	173,236	0	366	0	173,602	173,199	0	403	0
Kentucky	214,861	214,642	219	0	0	214,861	214,642	219	0	0
Louisiana	592,869	455,869	137,000	0	—	592,869	434,237	137,960	20,672	—
Maine	960,583	796,394	164,189	0	0	960,583	911,493	49,090	0	0
Maryland	21,002	17,255	3,641	106	0	21,001	20,999	0	2	0
Massachusetts	64,018	37,706	569	25,663	80	64,017	44,893	17,952	1,080	92
Michigan	489,433	472,587	—	16,846	0	489,433	489,433	—	0	0
Minnesota	2,190,320	225,838	0	1,964,482	0	1,536,185	1,290,496	68,088	177,601	0
Mississippi	500,000	339,358	160,596	46	0	500,000	492,453	7,501	46	0
Missouri	287,543	285,074	—	2,469	0	287,543	260,376	—	79	27,088
Montana	632,705	226,475	392,977	13,253	0	632,704	288,934	332,619	11,151	0
Nevada	308,125	149,042	159,083	0	0	308,125	43,075	264,100	950	0
New Hampshire	153,516	149,776	3,270	470	—	155,463	146,442	5,293	3,728	—
New Mexico	115,598	115,598	—	0	0	115,598	115,598	—	0	0
New York	750,000	560,000	—	190,000	0	750,000	678,000	—	72,000	0
North Carolina	305,345	295,221	—	10,124	—	305,355	303,134	—	2,221	—
North Dakota	611,073	367,484	21,855	221,734	—	610,749	376,767	12,248	221,734	—
Ohio	26,015	24,879	865	271	—	102,731	20,953	75,722	6,056	—
Oklahoma	448,776	219,855	200,493	28,428	0	434,431	285,197	123,310	25,924	0
Oregon	504,928	504,928	0	0	0	504,928	504,928	0	0	0
Puerto Rico	11,146	7,627	1,124	2,395	0	11,146	6,791	1,960	2,395	0
Rhode Island	16,683	15,553	100	487	543	16,683	15,553	100	487	543
South Carolina	260,902	175,069	72,713	13,120	0	217,667	197,142	20,525	0	0
South Dakota	682,034	682,034	—	0	0	682,034	682,034	—	0	0
Tennessee	540,030	483,675	—	56,355	—	538,322	521,473	—	16,849	—
Texas	1,543,898	1,543,884	—	14	—	1,543,898	1,536,346	—	7,552	—
Utah	133,754	0	133,754	0	0	133,754	2,329	131,425	0	0
Vermont	225,740	221,525	1,877	1,957	381	41,820	40,636	331	838	15
Virginia	160,640	150,158	10,482	0	—	160,640	160,640	—	0	—
Washington	182,848	62,551	5,277	115,020	—	182,807	180,512	660	1,635	—
West Virginia	18,870	12,988	4,152	1,730	0	18,870	18,870	0	0	0
Wisconsin	139,330	42,284	89,918	7,128	0	139,330	42,284	89,918	7,128	0
Wyoming	427,219	427,219	0	0	0	245,311	245,311	0	0	0
Totals	17,633,217	12,373,785	1,912,545	3,340,611	6,276	16,209,550	13,377,843	1,640,804	1,121,883	69,020
Percent of Assessed Waters		70.2%	10.8%	18.9%	0.0%		82.5%	10.1%	6.9%	0.4%

— Not reported

Source: U.S. Environmental Protection Agency, *National Water Quality Inventory*, 1990 Report to Congress, Apr. 1992, p. 26.

Designated Use Support in Estuaries

State	Estuary Square Miles	Square Miles Assessed: Total	Square Miles Assessed: Percent Evaluated	Square Miles Assessed: Percent Monitored	Sq. Miles Fully Supporting	Sq. Miles Threatened	Sq. Miles Partially Supporting	Sq. Miles Not Supporting
Alabama	625	103	12	88	65	35	3	0
California	1,598	799	26	74	267	—	39	493
Connecticut	601	601	4	96	363	4	229	5
Delaware River Basin	866	866	—	—	0	855	0	11
District of Columbia	6	6	0	100	0	0	0	6
Florida	4,298	2,712	27	73	1,639	25	861	187
Georgia	594	594	83	17	584	—	8	2
Hawaii	134	134	0	100	40	0	94	0
Louisiana	7,656	5,445	86	14	1,862	1,953	1,630	0
Maine	1,633	1,633	88	12	1,473	0	36	124
Maryland	2,523	2,523	15	85	109	10	2,312	92
Massachusetts	177	136	18	82	18	0	17	101
Mississippi	133	133	90	10	120	6	7	0
New York	1,564	1,564	0	100	1,140	11	150	263
North Carolina	3,194	3,194	0	100	2,894	8	290	2
Oregon	206	61	0	100	4	—	57	0
Rhode Island	193	193	26	74	142	17	18	16
South Carolina	2,155	364	0	100	244	0	2	118
Texas	1,990	1,990	0	100	1,505	0	0	485
Virginia	2,529	2,529	25	75	2,158	—	368	3
Virgin Islands	6	6	0	100	4	1	1	1
Washington	2,943	1,106	7	93	373	127	451	155
Totals	35,624	26,692	66	34	15,004	3,052	6,573	2,064
Percent of Assessed Waters					56.2%	11.4%	24.6%	8.0%

— Not reported.

Source: 1990 State Section 305(b) reports.

Impaired Estuary Square Miles Affected by Causes of Pollution

State	Total Impaired Waters*	Nutrients		Organic Enrichment		Pathogens		Priority Organics		Suspended Solids	
		Major	Mod/Min	Major	Mod/Min	Major	Mod/Min	Major	Mod/Min	Major	Mod/Min
District of Columbia	6	—	1	1	—	1	5	—	—	—	—
Florida	1,048	216	583	1	40	—	45	—	44	68	362
Georgia	10	—	—	2	4	—	—	—	—	—	—
Hawaii	94	—	46	—	—	—	—	—	—	—	—
Maine	160	—	—	1	—	38	—	3	—	—	—
Maryland	2,404	1,033	1,159	957	283	87	113	—	—	—	—
Massachusetts	118	1	14	2	16	95	16	15	18	—	28
New Jersey	143	70	—	—	—	142	—	70	—	—	—
New York	413	—	102	12	101	244	123	145	268	—	—
North Carolina**	292	—	—	226	—	35	—	—	—	—	—
Oregon**	57	—	—	—	—	57	—	—	—	—	—
Rhode Island	34	—	14	7	—	28	2	—	—	—	—
South Carolina	120	24	13	24	13	112	—	—	—	—	—
Texas	485	—	—	—	—	329	22	—	—	—	—
Virgin Islands	2	—	1	—	1	—	1	—	—	—	1
Washington	605	2	0	152	33	209	77	56	298	0	8
Totals	5,991	1,346	1,933	1,385	491	1,377	404	289	628	68	399
Combined Totals			3,279		1,876		1,781		917		467
Percent of Impaired Waters			54.7%		31.3%		29.7%		15.3%		7.8%

* The sum of partially and nonsupporting estuary square miles (Table 4-1).

** These States did not specify the degree of impact (i.e., Major or Moderate/Minor); estuary square miles were placed in the "Major" column for national reporting purposes.

— Zero or not reported.

Source: U.S. Environmental Protection Agency, *National Water Quality Inventory*, 1990 Report to Congress, Apr. 1992, p. 48.

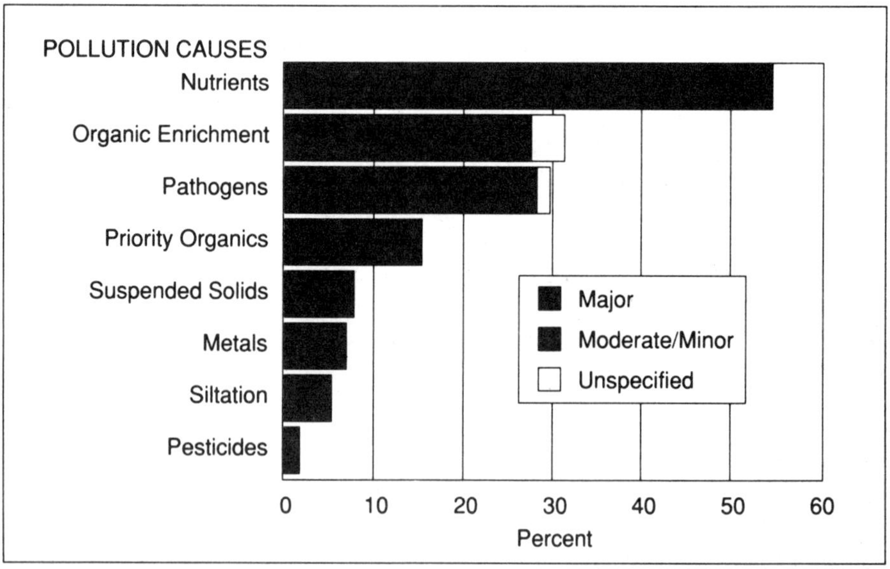

Percent of Impaired Estuary Square Miles Affected by Causes of Pollution

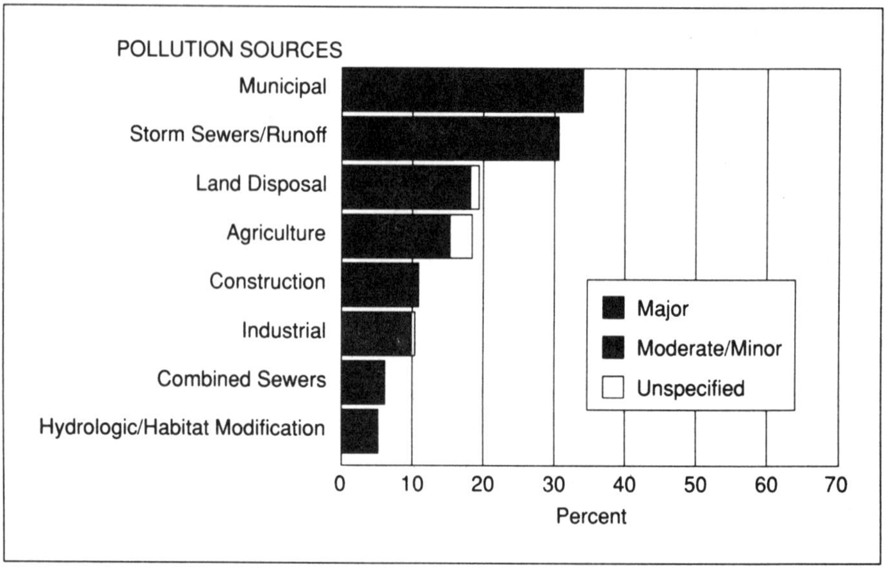

Percent of Impaired Estuary Square Miles Affected by Sources of Pollution

Source: U.S. Environmental Protection Agency, *National Water Quality Inventory*, 1990 Report to Congress, Apr. 1992, pp. 25–26.

Attainment of Clean Water Act Goals in Assessed Estuaries

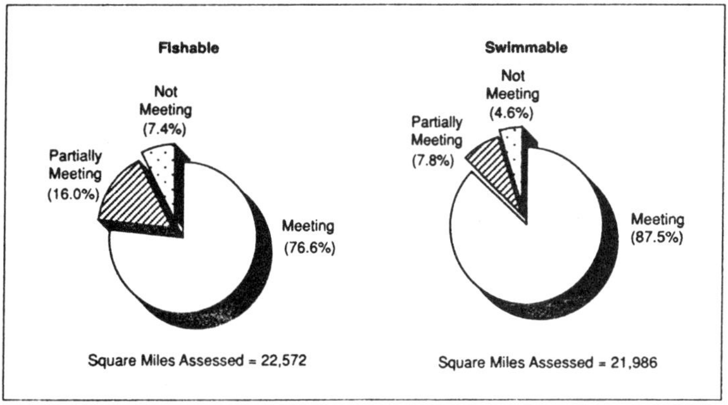

Attainment of Clean Water Act Goals in Estuaries

	Fishable Goal (square miles)					Swimmable Goal (square miles)				
State	**Assessed**	**Meeting**	**Partially Meeting**	**Not Meeting**	**Not Attainable**	**Assessed**	**Meeting**	**Partially Meeting**	**Not Meeting**	**Not Attainable**
Alabama	103	100	3	0	0	103	100	3	0	0
California	799	306	2	491	—	795	349	0	446	—
Connecticut	600	366	218	16	0	600	554	16	30	0
Delaware River Basin	866	802	4	60	0	866	855	0	11	0
District of Columbia	6	0	0	6	0	6	0	0	6	0
Florida	2,730	2,410	133	187	0	2,730	2,410	133	187	0
Georgia	594	584	8	2	0	0	—	—	—	—
Hawaii	40	40	—	0	0	40	40	—	0	0
Louisiana	4,684	4,432	252	0	—	4,685	3,535	1,078	72	—
Maine	1,633	1,475	36	122	0	1,633	1,631	0	2	0
Maryland	2,522	107	2,283	132	0	2,522	2,520	0	2	0
Massachusetts	137	78	13	46	0	137	53	46	38	0
Mississippi	133	126	7	0	0	133	132	1	0	0
Oregon	61	4	57	0	—	61	61	0	0	—
New York	1,569	1,229	—	335	5	1,578	1,487	—	77	14
Rhode Island	193	182	0	4	7	193	182	0	4	7
South Carolina	364	338	25	1	0	364	320	18	26	0
Texas	1,991	1,933	—	58	—	1,990	1,990	—	0	—
Virginia	2,529	2,158	368	3	—	2,529	2,158	368	3	—
Virgin Islands	6	4	1	1	0	5	3	1	1	0
Washington	1,012	605	210	197	—	1,016	860	49	107	—
Totals	22,572	17,279	3,620	1,661	12	21,986	19,240	1,713	1,012	21
Percent of Assessed Waters		76.6%	16.0%	7.4%	0.0%		87 5%	7 8%	4 6%	0.0%

— Not reported.

Source: U.S. Environmental Protection Agency, *National Water Quality Inventory*, 1990 Report to Congress, Apr. 1992, p. 54.

Designated Use Support in Oceans

State	Ocean Square Miles	Coastal Miles Assessed: Total	Coastal Miles Assessed: Percent Evaluated	Coastal Miles Assessed: Percent Monitored	Miles Fully Supporting	Miles Threatened	Miles Partially Supporting	Miles Not Supporting
Alabama	50	50	100	0	50	0	0	0
California	1,840	1,118	48	52	1,046	—	49	23
Delaware	25	25	100	0	25	0	0	0
Florida	8,460	880	37	63	832	0	36	12
Hawaii	824	824	62	38	824	0	0	0
Maryland	32	32	0	100	32	0	0	0
Mississippi	81	81	0	100	0	10	60	11
New York	130	130	0	100	60	0	70	0
Puerto Rico	434	434	72	28	252	38	75	69
Texas	367	367	0	100	367	0	0	0
Virginia	112	112	100	0	112	—	0	0
Virgin Islands	173	14	0	100	12	1	0	1
Washington	165	163	100	0	163	0	0	0
Totals	12,693	4,230	48	52	3,775	49	290	116
Percent of Assessed Waters					89.2%	1.2%	6.9%	2.7%

— Not reported.

Attainment of Clean Water Act Goals in Oceans

State	Fishable Goal (coastal miles): Assessed	Fishable: Meeting	Fishable: Partially Meeting	Fishable: Not Meeting	Fishable: Not Attainable	Swimmable Goal (coastal miles): Assessed	Swimmable: Meeting	Swimmable: Partially Meeting	Swimmable: Not Meeting	Swimmable: Not Attainable
Alabama	50	50	0	0	0	50	50	0	0	0
California	1,061	1,000	49	12	—	1,044	1,034	0	10	—
Delaware	25	25	0	0	0	25	25	0	0	0
Florida	892	861	19	12	0	892	861	19	12	0
Hawaii	824	824	—	0	0	824	824	—	0	0
Maryland	32	32	0	0	0	32	32	0	0	0
Mississippi	81	10	60	11	0	81	10	71	0	0
New York	130	60	—	70	0	130	130	—	0	0
Puerto Rico	434	319	88	23	4	434	238	60	41	95
Texas	367	367	—	0	—	367	367	—	0	—
Virgin Islands	13	12	0	1	0	11	10	0	1	0
Virginia	112	112	0	0	—	112	112	0	0	—
Washington	163	163	0	0	0	163	163	0	0	0
Totals	4,184	3,835	216	129	4	4,165	3,856	150	64	95
Percent of Assessed Waters		91.6%	5.2%	3.1%	0.0%		92.6%	3.6%	1.5%	2.2%

— Not reported.

Source: U.S. Environmental Protection Agency, *National Water Quality Inventory*, 1990 Report to Congress, Apr. 1992, pp. 55–58.

Wetlands Losses in the United States, 1780s to 1980s

State	Total Surface Area (acres)*	Wetlands: Estimates of Original Wetlands circa 1780s	Wetlands: Percent of Surface Area	Wetlands: Estimates of Existing Wetlands circa 1980s	Wetlands: Percent of Surface Area	Wetlands: Percent of Wetlands Lost
Alabama	33,029,760	7,567,600	22.9	3,783,800	11.5	50
Arizona	72,901,760	931,000	1.3	600,000	0.8	36
Arkansas	33,986,560	9,848,600	29.0	2,763,600	8.1	72
California	101,563,520	5,000,000	4.9	454,000	0.4	91
Colorado	66,718,720	2,000,000	3.0	1,000,000	1.5	50
Connecticut	3,205,760	670,000	20.9	172,500	5.4	74
Delaware	1,316,480	479,785	36.4	223,000	16.9	54
Florida	37,478,400	20,325,013	54.2	11,038,300	29.5	46
Georgia	37,680,640	6,843,200	18.2	5,298,200	14.1	23
Idaho	53,470,080	877,000	1.6	385,700	0.7	56
Illinois	36,096,000	8,212,000	22.8	1,254,500	3.5	85
Indiana	23,226,240	5,600,000	24.1	750,633	3.2	87
Iowa	36,025,600	4,000,000	11.1	421,900	1.2	89
Kansas	52,648,960	841,000	1.6	435,400	0.8	48
Kentucky	25,852,800	1,566,000	6.1	300,000	1.2	81
Louisiana	31,054,720	16,194,500	52.1	8,784,200	28.3	46
Maine	21,257,600	6,460,000	30.4	5,199,200	24.5	20
Maryland	6,769,280	1,650,000	24.4	440,000	6.5	73
Massachusetts	5,284,480	818,000	15.5	588,486	11.1	28
Michigan	37,258,240	11,200,000	30.1	5,583,400	15.0	50
Minnesota	53,803,520	15,070,000	28.0	8,700,000	16.2	42
Mississippi	30,538,240	9,872,000	32.3	4,067,000	13.3	59
Missouri	44,599,040	4,844,000	10.9	643,000	1.4	87
Montana	94,168,320	1,147,000	1.2	840,300	0.9	27
Nebraska	49,425,280	2,910,500	5.9	1,905,500	3.9	35
Nevada	70,745,600	487,350	0.7	236,350	0.3	52
New Hampshire	5,954,560	220,000	3.7	200,000	3.4	9
New Jersey	5,015,040	1,500,000	29.9	915,960	18.3	39
New Mexico	77,866,240	720,000	0.9	481,900	0.6	33
New York	31,728,640	2,562,000	8.1	1,025,000	3.2	60
North Carolina	33,655,040	11,089,500	33.0	5,689,500	16.9	49
North Dakota	45,225,600	4,927,500	10.9	2,490,000	5.5	49
Ohio	26,382,080	5,000,000	19.0	482,800	1.8	90
Oklahoma	44,748,160	2,842,600	6.4	949,700	2.1	67
Oregon	62,067,840	2,262,000	3.6	1,393,900	2.2	38
Pennsylvania	29,013,120	1,127,000	3.9	499,014	1.7	56
Rhode Island	776,960	102,690	13.2	65,154	8.4	37
South Carolina	19,875,200	6,414,000	32.3	4,659,000	23.4	27
South Dakota	49,310,080	2,735,100	5.5	1,780,000	3.6	35
Tennessee	27,036,160	1,937,000	7.2	787,000	2.9	59
Texas	171,096,960	15,999,700	9.4	7,612,412	4.4	52
Utah	54,346,240	802,000	1.5	558,000	1.0	30
Vermont	6,149,760	341,000	5.5	220,000	3.6	35
Virginia	26,122,880	1,849,000	7.1	1,074,613	4.1	42
Washington	43,642,880	1,350,000	3.1	938,000	2.1	31
West Virginia	15,475,840	134,000	0.9	102,000	0.7	24
Wisconsin	35,938,560	9,800,000	27.3	5,331,392	14.8	46
Wyoming	62,664,960	2,000,000	3.2	1,250,000	2.0	38
Subtotal (Conterminous U.S.)	1,934,198,400	221,129,638	11	104,374,314	5	53
Alaska	375,303,680	170,200,000	45.3	170,000,000	45.3	0.1
Hawaii	4,115,200	58,800	1.4	51,800	1.3	12
Total U.S.	2,313,617,280	391,388,438		274,426,114	11.9	30

*Note: Surface area includes both land and water surface area. There are some discrepancies between the total surface area of States. These differences are probably due to shifting river channels forming State borders. The area given is that presented by the U.S. Geological Survey, National Atlas of the United States, 1970.

Source: U.S. Environmental Protection Agency, *National Water Quality Inventory*, 1990 Report to Congress, Apr. 1992, p. 70.

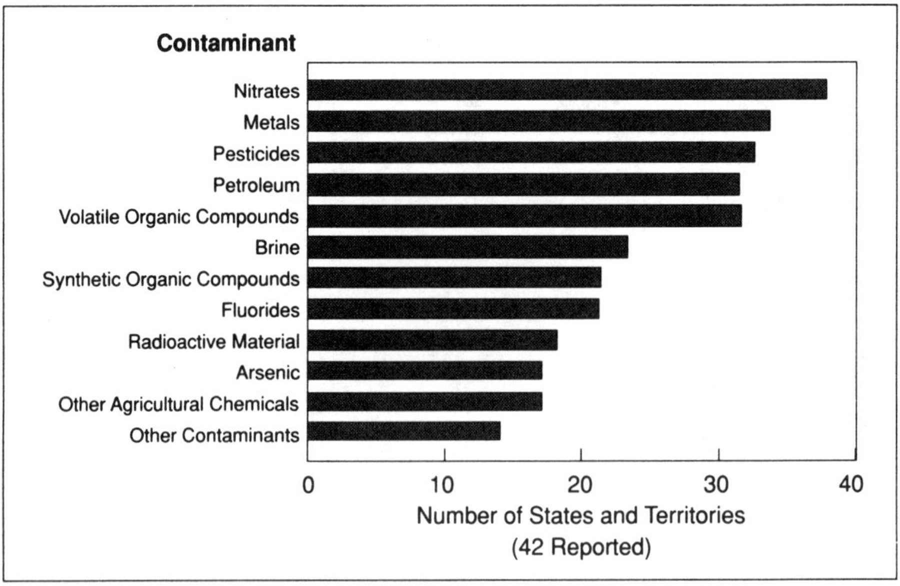

Most Frequently Observed Ground-Water Contaminants by Number of States and Territories Reporting

Source: U.S. Environmental Protection Agency, *National Water Quality Inventory*, 1990 Report to Congress, Apr. 1992, p. 116.

APPENDIX C

HAZARDOUS AND SOLID WASTE CHARTS

Superfund Remediation at National Priority List Sites 328

National Priority List Sites 329

Predominant Waste Types at NPL Sites, 1991 330

Environmental Damage at NPL Sites, 1991 330

Municipal Solid Waste, 1960-1988 331

Municipal Solid Waste Generation and Recovery, by Waste Type, 1960-1988 331

Superfund remediation at National Priority List sites.

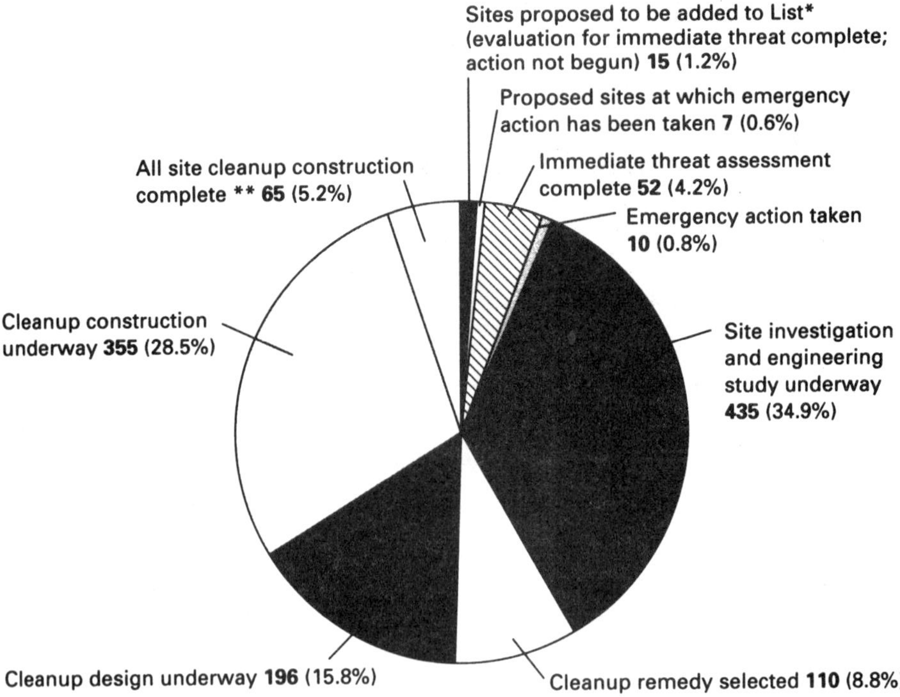

* The Environmental Protection Agency (EPA), which administers Superfund, has asked for comments on its plans to add these sites to the National Priorities List (NPL), making them eligible for permanent Superfund cleanup.

** As of February 10, 1992, all cleanup construction had been completed at 70 sites.

Note: Segments are cumulative; for example, sites with construction underway have already completed study, remedy selection, and design. As of December 31, 1991, of the 1,205 NPL sites or proposed sites or the 40 sites deleted from the List, 119 were located at federal facilities. The cleanup of these sites is the primary responsibility of the departments and agencies that operate them; by law, Superfund cannot pay for their permanent cleanup.

Source: The Council on Environmental Quality, 22nd Annual Report, Mar. 1992, p. 112.

National Priority List sites.

o = Final site
* = Proposed site

Not shown: Alaska, 6; Guam, 1; Hawaii, 2; and Puerto Rico, 9; as of September 10, 1991.

National totals: Final NPL sites 1,185; Proposed NPL sites 22; Total NPL sites 1,207

Source: The Council on Environmental Quality, 22nd Annual Report, Mar. 1992, p. 113.

Predominant waste types at NPL sites, 1991.

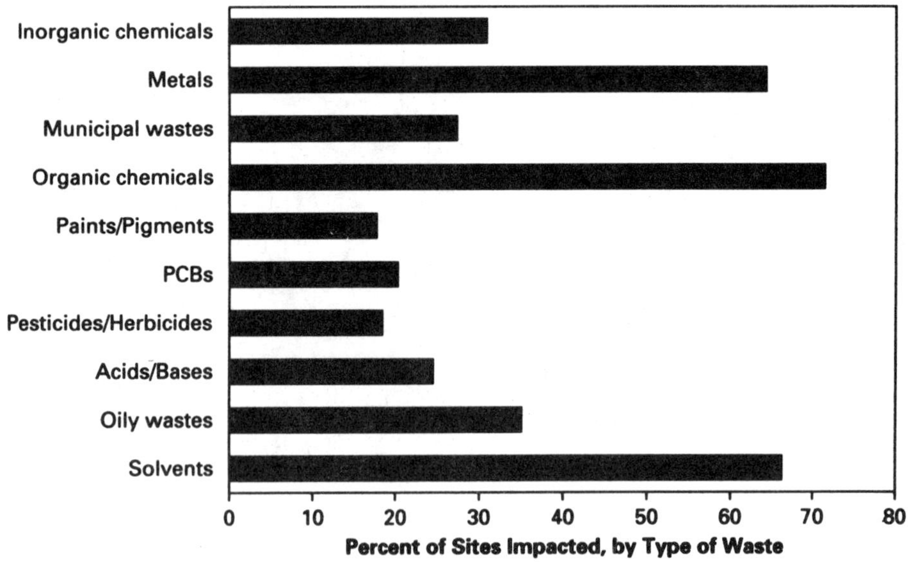

Note: NPL=National Priority List. PCBs=Polychlorinated biphenyls.

Environmental damage at NPL sites, 1991.

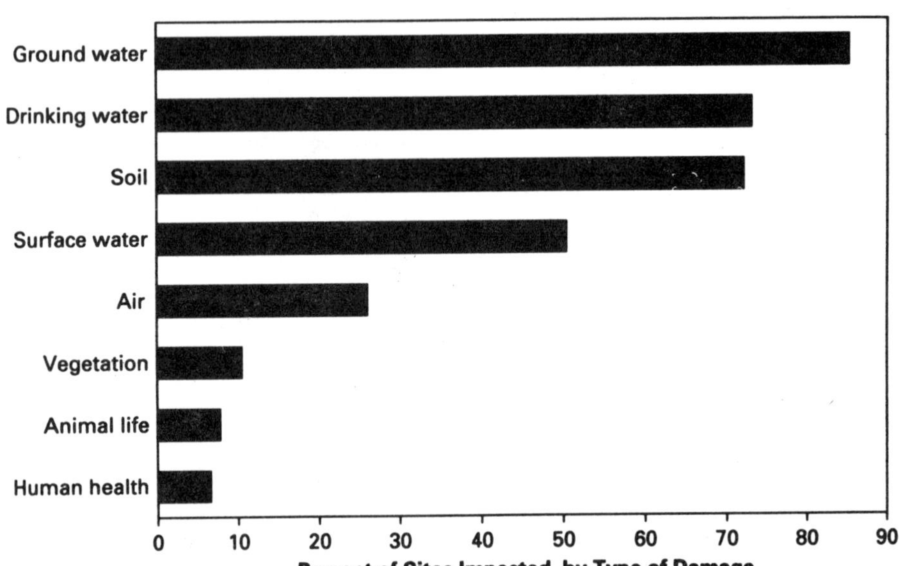

Note: NPL=National Priority List.

Source: The Council on Environmental Quality, 22nd Annual Report, Mar. 1992, p. 334.

Municipal solid waste, 1960-1988.

Year	Gross discards		Materials recovery			Net discards	
	Total	Per capita	Total	Per capita	Recovery	Total	Per capita
	million tons	*pounds per day*	*million tons*	*pounds per day*	*%*	*million tons*	*pounds per day*
1960	87.8	2.66	5.9	0.18	6.7	81.9	2.48
1965	103.4	2.92	6.8	0.19	6.6	96.6	2.73
1970	121.9	3.27	8.6	0.23	7.1	113.3	3.04
1975	128.1	3.26	9.9	0.25	7.7	118.2	3.01
1980	149.6	3.61	14.5	0.35	9.7	135.1	3.26
1985	161.6	3.70	16.4	0.38	10.1	145.2	3.32
1988	179.6	4.00	23.5	0.52	13.1	156.0	3.48

Notes: Totals may not agree with detail because of independent rounding.

Municipal solid waste generation and recovery, by waste type, 1960-1988.

Year	Paper		Glass		Metal		Aluminum		Plastics	
	Generation	Recovery	Generation	Recovery	Generation	Recovery	Generation	Recovery	Generation	Recovery
	million tons									
1960	29.9	5.4	6.7	0.1	10.1	0.1	0.4	0.0	0.4	0.0
1965	38.0	5.7	8.7	0.1	10.6	0.4	0.5	0.0	1.4	0.0
1970	44.2	7.4	12.7	0.2	13.3	0.4	0.8	0.0	3.1	0.0
1975	43.0	8.2	13.5	0.4	13.2	0.6	1.1	0.1	4.5	0.0
1980	54.7	11.9	15.0	0.8	12.7	0.9	1.8	0.3	7.8	0.0
1985	61.5	13.1	13.2	1.0	11.9	0.6	2.3	0.6	11.6	0.1
1988	71.8	18.4	12.5	1.5	12.7	1.4	2.5	0.8	14.4	0.2

Year	Rubber & Leather		Textiles		Wood		Food		Yard	
	Generation	Recovery	Generation	Recovery	Generation	Recovery	Generation	Recovery	Generation	Recovery
	million tons									
1960	2.0	0.3	1.7	0.0	3.0	0.0	12.2	0.0	20.0	0.0
1965	2.6	0.3	1.9	0.0	3.5	0.0	12.7	0.0	21.6	0.0
1970	3.2	0.3	2.0	0.0	4.0	0.0	12.8	0.0	23.2	0.0
1975	3.9	0.2	2.2	0.0	4.4	0.0	13.4	0.0	25.2	0.0
1980	4.3	0.1	2.6	0.0	4.9	0.0	13.2	0.0	27.5	0.0
1985	3.8	0.2	2.8	0.0	5.4	0.0	13.2	0.0	30.0	0.0
1988	4.6	0.1	3.9	0.0	6.5	0.0	13.2	0.0	31.6	0.5

Source: The Council on Environmental Quality, 22nd Annual Report, Mar. 1992, pp. 335–336.

INDEX

A

Acid rain 123, 143–147
Administrative orders 15, 271–272
Administrative proceedings (*see also* Hearings)
  FIFRA cases 101–102
  hazardous materials cases, examples 219–222
  TSCA cases 49–51
Adverse effects rule
  records 24–25
  reports 24–25, 82–83
Air pollutants
  hazardous pollutants 147–153
  toxic pollutants 123
Air quality standards (*see also* National ambient air quality standards (NAAQS) 7
Air stripping 182
Air toxics 147–149
Air transport 208–209
Annual report 80–81
Application process
  experimental use permits 87
  pesticide registrations 74–77, 89–90
Applicator certification of pesticides 69–72
Army Corps of Engineers (Corps), U.S. 161
Arsenic 150
Asbestos 43–44, 150–151
Attorney general 15–16

B

Benzene 149, 151
Beryllium 151–152
Best available technology economically achievable (BAT) 166–167
Best conventional pollution control technology (BCT) 166–167
Best practicable control technology currently available (BPT) 166–167
Biotechnology products 39

C

Carbon monoxide 126
Cash compensation for cancelled or suspended products 93
Certifications
  hazardous materials shipments 203–204
  imports 46–47
  pesticides 69–72

CFCs 42–43
Chemicals
  exemptions 117
  exports 45–47
  hazardous chemicals 39–45
  imports 45–47
  inventories 106, 112–115
  new chemicals 30, 32–39
  recordkeeping requirements 19–26
  reporting requirements 19–26, 116–118
  testing 26–32
  toxic chemical release reporting 115–119
Child-resistant packaging 73–74
Chromium 44
Citizens' suits
  under Clean Water Act 178–179
  under RCRA 257–258
  under TSCA 47–48
Civil penalties 15
  under CERCLA 271–272
  EPA enforcement authority 175, 256
  under FIFRA 99–100
  under HMTA 217
  RCRA tank requirements 239
  Safe Drinking Water Act 185
  under TSCA 50–51
Classifications
  DOT's Hazardous Materials Table 192
  hazardous materials 188, 191–193
  PCB equipment reclassification 41
  pesticides 68–69
Clean Air Act 6–8
  1970 Amendments 1
  1990 Amendments 123–157
Clean Water Act 6, 8–9, 158–179
Coast Guard, U.S.
  administrative proceedings, examples 221
  HMTA, enforcement and compliance monitoring 218–219
  water transport responsibility 207
Color standards 59
Comprehensive assessment information rule (CAIR) 19, 21
Comprehensive Environmental Response, Compensation, and Liability Act (CERCLA) 2–3, 12, 109, 222, 258–282, 284
Confidentiality
  business information 22
  Statement of Formula 66
Containers
  hazardous materials 196–197
  hazardous waste 231
  pesticides 283–284
Criminal penalties 15–16
  EPA enforcement authority 175, 256
  Safe Drinking Water Act 185
  under FIFRA 99–100
  under HMTA 217

D

Data Call–In (DCI) program 61–62
Data requirements
  pesticides 64–68
  tables 59
Defenses 277–279
Department of (*see* specific departments)
Discharge permits (*see also* National Pollutant Discharge Elimination System (NPDES) 159, 171
Disposal and clean-up 245–258, 282–284
Dumps, state plans 227

E

Effluent limitations 162–166
Emergencies
  hazardous materials spills 211–214
  pesticides, exemptions 78–80
  Superfund removal actions 266–268
Emergency Planning and Community Right-to-Know Act (EPCRA) 105–122, 212, 243–244
Emissions
  evaporative 141–142
  exhaust 139–141
  limits, EPA authority 7
Environmental Protection Agency (EPA)
  administrative orders 271–272
  air quality standards 7
  authority 1
  Clean Air Act administration 5
  Clean Water Act administration 8
  enforcement responsibility 101–102, 122, 124, 155–158, 175–179, 184–185, 239, 255–258, 270–271
  Federal Insecticide, Fungicide, and Rodenticide Act administration 5
  on hazardous air pollutants 147–149
  hazardous materials storage, authority for 240
  on hazardous substances 212
  hazardous waste regulation 11
  municipal solid waste landfills standards 226
  oil, spill response and cleanup responsibility 212
  storage tanks 12, 233–238
  Toxic Substances Control Act administration 4
  underground storage tanks 12, 233–238
Equipment reclassification, PCB 41
Evaporative emissions 141–142
Exhaust emissions 139–141
Experimental use permits 86–87

Exports
of chemicals 45–47
of pesticides 94–95

F

Facilities (*see also* Treatment, storage, and disposal (TSD) facilities)
defined 107–108
notice requirements 107–108
NSPS requirements 130–133, 137
reporting requirements 107–108, 115
Federal Aviation Administration
administrative proceedings, examples 220
as mode-specific agency 207
HMTA, enforcement and compliance monitoring 218–219
Federal Highway Administration
administrative proceedings, examples 220
as mode-specific agency 207
HMTA, enforcement and compliance monitoring 218–219
Federal Insecticide, Fungicide, and Rodenticide Act (FIFRA) 2, 5–6, 52–104, 282–284
Federal Railroad Administration
administrative proceedings, examples 219–220
as mode-specific agency 207
HMTA, enforcement and compliance monitoring 11, 218–219
Federal Water Pollution Control Act Amendments of 1971 2
Fees
pesticide registration program 68
premarket notification process 34–35
Fire rules, PCB 41
Formulator Exemption Statement 66

G

Generators 229–230
Good laboratory practices (GLPs) 63–64

H

Hazard Communication Standard
applicable substances 112
revisions 191
Hazard Ranking System (HRS) 260–261
Hazardous air pollutants 147–153
Hazardous chemicals 39–45
Hazardous materials
marking requirements for shipments 197–199
storage under TSCA 240–242
Hazardous Materials Transportation Act (HMTA) 10–11, 187–222
Hazardous Substances and Reportable Quantities (RQ) list 193, 212
Hazardous waste
disposal 245–258
EPA regulation 11–12
identification 227–229
manifest 201, 204, 231
reporting requirements 213
storage 239–240
transport of 231–232
Health and safety studies, unpublished 25
Hearings 15, 93
Highway transport of hazardous materials 210–211

I

Identifications
hazardous materials shipments 197–199
hazardous substances 261–262
hazardous waste containers 231
Imminent hazards 40
Imports
of chemicals 45–47
of pesticides 94
Indemnification for cancelled or suspended products 93
Industrial sewage treatment 164–166
Injunctions 15
Inspections
hazardous waste facilities 256
PCB–containing equipment 42
pesticides 80–84
product registration data 67–68
TSCA violations 48
Inventories
annual chemical inventories 112–115
toxic substances 22–24

J

Justice Department, U.S. (*see also* Attorney general)
FIFRA violations 103
right to filc civil action 175–176

L

Labeling
hazardous materials shipments 190, 199–200
PCB 41
pesticides 72–74, 94–96
Land disposal of waste, ban on 253–254
Landfills, EPA standards 226

Lead contamination
air pollutant 126
drinking water 183

M

Marking requirements for hazardous materials shipments 197–199
Mercury 152
Metal-working fluids 44–45
Motor vehicles (*see* Recalls, Emissions)
Municipal sewage treatment 167–168

N

National ambient air quality standards (NAAQS) 123–125
National Contingency Plan (NCP) 12
National Pollutant Discharge Elimination System (NPDES) 6, 8, 123, 159
National Response Center 211–212
Navigable waters
defined 160–161, 170
U.S. Army Corps of Engineers (Corps) authority 161
Negligence 14–15
Negotiated agreements 29
Nitrogen dioxide 127
Nixon, Richard 1–2
Notice requirements
under EPCRA 108, 243–244
export of chemicals 45–46
hazardous materials shipments 191
storage tanks 234–235, 243–244
substantial risk 25–26
NPES permits 159, 171
NSPS requirements 130–133, 137
Nuisance 14

O

Open dumps, state plans 227
Ozone 127–128

P

Packaging requirements
child-resistant packaging 73–74
hazardous materials shipments 190, 194–197
hazardous waste 231
pesticides 96
Particulates 127–128
Parties in CERCLA suits 273–274
PCBs 40–42, 240–242
Penalties (*see* Civil penalties, Criminal penalties)
Permits
air pollutants covered by NAAQS and NESHAP 155
discharge permits 123, 159
emergency-use permits 78
experimental use permits 87
NPDES permits 249–253
stormwater 172–174
TSD facilities 249–253
Pesticides (*see also* Federal Insecticide, Fungicide, and Rodenticide Act (FIFRA)
applicator certifications 69–72
assessment guidelines 62–63
definition 53
disposal 282–284
minor-use pesticides 61
registration, exemptions 58
reregistration of older pesticides 87–90
risky pesticides 90–94
storage 242–243
suspension of pesticides 91
Petitions
pesticide tolerance petition process 85–86
under TSCA 47–48
Placards for hazardous materials shipments 200–201
Place of business registration 80–81
Pollution (*see also* Air pollutants, Pollution Prevention Act
controls 2, 6–12
nonpoint source pollution 172–175
prevention 6–12
Pollution Prevention Act 119
Pre-emption of local pesticide regulation 96
Preliminary assessment information rule (PAIR) 19–21
Premarker notification process (PMN)
exemptions 33–34
fees 34–35
forms 34–35
recordkeeping 34–35
Product registration of pesticides 55–68
PSD programs 137
Public access (*see* Inspections)
Public health exemptions 78–80

Q

Quarantine, pesticides 78–80

R

Radionuclides 152
Radon-222 153
Rail transport of hazardous materials 207

Recalls
  canceled and suspended products 283–284
  motor vehicles 142–143
Recordkeeping requirements
  chemical regulation 19–26
  hazardous waste transport 231
  PCB-containing equipment 42
  pesticides 80–84
  premarket notification process 34–35
Recycling 11, 222, 225
Registrations
  cancellations 92–93
  establishment registrations 80–81
  older pesticides, reregistrations 87–90
  pesticides 55–68
  place of business 80
  review standards 74–77
  special needs registrations 96–97
Releases
  chemicals 109–112
  hazardous substances 109–112
  hazardous waste 274
Reporting requirements
  CERCLA 109–110
  chemical regulation 19–26, 116
  enforcement 121–122
  EPCRA 112
  hazardous materials or waste, accidental discharge 213, 232
  hazardous substances 212
  inventory reporting 114
  MSDS reporting 113
  pollution prevention 119
  public access to reports 114–115
  substances and facilities covered 107–108
  toxic chemcial release reporting 115–116
Resource Conservation and Recovery Act (RCRA) 2, 6, 11–12, 222–240, 245–258, 284
Reverse osmosis 182
Right–to–know requirements (*see* Emergency Planning and Community Right–to–Know Act (EPCRA)
RSPA
  administrative proceedings, examples 221–222
  HMTA, enforcement and compliance monitoring 218–219

S

Safe Drinking Water Act (SDWA) 9–10, 179–185, 284–285
Safety and health studies, unpublished 25
Section 107 enforcement suits 273
Section 5(e) orders 35–36
Settlement agreements 279–282
Sewage
  combined systems 174–175
  industrial treatment 164–166
  municipal treatment 167–168
  separate systems 174–175
Shippers and shipping
  certifications 203–204
  defined 190
  papers 201
  requirements 190–191
Significant new rules (SNUR) 37–38
Solid waste
  definition 224–225
  exclusions 225
  nonhazardous solid waste 225–226
  under RCRA 11–12
  state plans 226–227
Spills, emergency response 211
State responsibilities
  air quality, implementation plans 133–135
  authority under FIFRA 95–97
  CERCLA 269
  enforcement responsibility under FIFRA 97–104
  hazardous materials, transport of 215–217
  hazardous waste programs 254–255
Statute of limitations 49–50
Steam stripping 182
Storage (*see also* Underground storage tanks (UST))
  hazardous waste 239–242
  pesticides 242–243
Stormwater permits 172–174
Subpoenas, TSCA violations 48–49
Substantial risk
  notices 25–26
  reporting 83–84
Sulfur dioxide 130
Superfund (*see* Comprehensive Environmental, Response, Compensation, and Liability Act (CERCLA)
Suspension of pesticides 91

T

Testing of chemicals 26–32
Tolerances for pesticides 84–87
Toxic air pollutants 123
Toxic Release Inventory (TRI) 106
Toxic Substances Control Act (TSCA) 2, 4–5, 17–52, 240–242
Trade secrets
  chemicals, reporting requirements 119–121
  pesticides 67
Transportation Department (DOT) 10, 187–232

Treatment, storage, and disposal (TSD) facilities 246–253
Trespass 14

U

Underground storage tanks (UST)
  financial requirements 237–238
  RCRA regulation 11–12, 233–238
  state programs 238
  technical requirements 235–236
  trust fund 239
United nations 215
Urban ozone 123
U.S. Army Corps of Engineers (Corps) 161
U.S. Coast Guard
  administrative proceedings, examples 221
  HMTA, enforcement and compliance monitoring 218–219
  water transport, responsibility 207
U.S. Justice Department (*see also* Attorney general)
  FIFRA violations 103
  right to file civil action 175–176

V

Vinyl Chloride 153

W

Warnings
  hazardous waste containers 231
  pesticide labels 73
Water quality 161–162
Water transport of hazardous materials 209–210
Wetlands 170–171

ABOUT THE AUTHORS

James A. Stimson is editor of BNA's *Environment Reporter*. Mr. Stimson was previously editor of Education Funding News and Government Information Services, and was also a writer for Lansing State Journal. He holds a Bachelor of Science from the University of Michigan.

Jeffrey J. Kimmel heads an environmental consulting business in Brookfield, Vermont, and was formerly a legal editor with BNA's *Environment Reporter*. He received his J.D. and a Masters in Environmental Law from the Vermont Law School. Mr. Kimmel prepared the 1992 Edition of BNA Books' *Spill Reporting Procedures Guide*.

Sara Thurin Rollin is a reporter for BNA's *Chemical Regulation Reporter* and a regular contributor to BNA's *Daily Environment Report.* Her specialty is U.S. industrial chemical regulation, but she has also written about other occupational safety and environmental law issues. Ms. Rollin holds an M.S. in natural resource policy and a B.S. in national resource economics.